Memoirs of German Pilots in the First World War

Volume 2

Rosenstein, Böhme, and Schaefer

Translated and Introduced by Jason Crouthamel

Acknowledgements

For our aviation books, please see our website at: www.aeronautbooks.com.

I am looking for photographs of the less well-known German aircraft of WWI to complete this series. For questions or to help with photographs you may contact me at jherris@me.com.

Interested in WWI aviation? Join The League of WWI Aviation Historians (**www.overthefront.com**), Cross & Cockade International (**www.crossandcockade.com**), and Das Propellerblatt (**www.propellerblatt.de**).

ISBN: 978-1-953201-50-8
© 2022 Aeronaut Books, all rights reserved
Text © 2022 Jason Crouthamel
Color Profiles: Jim Miller
Design and layout: Jack Herris
Cover design: Aaron Weaver
Digital photo editing: Jack Herris

Table of Contents

Introduction	3
Willy Rosenstein	6
Erwin Böhme	34
Emil Schaefer	92

Left: Willy Rosenstein.

Introduction

This is the second volume to *Memoirs of German Pilots in the First World War*. In volume one I provided a more extensive introduction on the background and significance of these memoirs. The themes that I explored there on the psychological stress of aerial combat, ideals of masculinity and nationalism, notions of comradeship, and the different ways that pilots narrated their experiences with technology and the sensation of flight are also relevant to the three memoirs selected here by Willy Rosenstein, Erwin Böhme, and Emil Schaefer. At the same time, this volume's set of narratives offer distinct insights into the experiences of German pilots in the Great War. Their memoirs provide more than a glimpse into the experiences of front-line pilots. They also shed light on the very personal world of men and women trying to survive total war (as is the case with Böhme), and they illuminate the postwar crises that engulfed German history (as is the case with Rosenstein).

These memoirs by Rosenstein, Böhme and Schaefer are dramatically different kinds of narratives that were produced under very different circumstances. Rosenstein's autobiography was never published. It was typed by him in a South African internment camp in 1940. He had fled to South Africa in 1936 to escape persecution in Nazi Germany, and he was briefly imprisoned at 'Internment Camp No.2' in Ganspan, along with other former German citizens, as an 'enemy alien' after World War II broke out. This was bitterly ironic, as he had just been attacked for being Jewish in his native Germany, where Nazi racial laws defined him as an inferior, non-citizen. His "Autobiographical Note" typed in the South African internment camp is both a plea for freedom and an attempt to come to terms with how his own German nation, for which he had fought and sacrificed as a pilot in the First World War, could have turned against him as a racial 'outsider.'

Since Rosenstein did not produce his autobiography for mass consumption, it documents many of the hidden costs of war that were left out of wartime memoirs, which were designed to prop up a mythical image of pilots. Postwar problems including financial difficulties, failed marriages, drug addiction, and political crisis are on full display in Rosenstein's narrative. These problems afflicted many pilots who survived the war (Udet, for example). Yet Rosenstein's particular background as a German-Jewish pilot makes his narrative exceptional and rather extraordinary. Rosenstein flew in *Jasta* 27 in late 1917 as the wingman of Hermann Göring, the future head of the Gestapo and leading figure in the Nazi regime. Göring's anti-Semitism, which would culminate in his work as one of the architects of the extermination of the Jews during World War II, was already part of his world-view in the First World War. At a gathering in their squadron mess, Göring directed an anti-Semitic insult at his German-Jewish comrade. This would be a decisive moment in Rosenstein's life. Rosenstein escaped Göring's bigotry in 1917 to find acceptance in another squadron where his comrades saw him as he wanted to be seen, first and foremost as a German. His hopes for respect were buttressed by a lifelong friendship with his new commander Carl Degelow and other comrades who eventually helped him escape Germany shortly after the Nazis came to power in 1933. Despair at the loss of his wife, torment from friends who turned out to be Nazis, a harrowing odyssey to South Africa, and spiraling drug addiction, nearly shattered Rosenstein. But incredibly, he would eventually find refuge in Rustenberg, where he built an airfield near his farm and trained pilots. One of these pilots included his son, Ernest, who would join the South African Air Force and fly with the RAF and die fighting against Nazi Germany. The horrible irony must not have been lost on Rosenstein, who had to flee from, and then send his son to war against, the nation that he had so patriotically defended in 1914–1918.

If Rosenstein's autobiography gives us a glimpse into the hidden traumas experienced by men after the war, Erwin Böhme's letters shed light on a life usually concealed from public view, but from a different angle. Though Böhme maintained the image of a war hero who selflessly sacrificed for the nation, his letters, which were published after the war as *Letters from a German Fighter Pilot to a Young Girl*, suggest a much more complex, and even dissonant, inner life. Böhme's correspondence with his fiancée, Annamarie Brüning, reveal how many individuals experienced extreme dichotomies in war. Brutalized by violence, they also experienced the intensity of love and compassion needed to cope with stress. The two sides of this life experience is rarely exposed to historians, but through the intimacy of letter-writing, it's possible to reconstruct this very secret world that two individuals never imagined would be read by us a hundred years later.

Böhme's letters offer a multi-dimensional picture of his life at the front, where he maintained his image and emulated the self-sacrifice and discipline of his heroes and friends, Oswald Bölcke and Manfred von Richthofen. At the same time, behind this image was growing disenchantment with the war. Böhme felt comfortable sharing this other side of himself with his fiancée. His correspondence with Annamarie reveals how, despite a widening experiential gulf between soldiers on the combat front and civilians at home, men and women still relied on each other for emotional support. Böhme shared with Annamarie his despair over the loss of Oswald Bölcke in a mid-air collision. The sense of guilt and helplessness that Böhme felt, evidence that he was psychologically traumatized by the loss of friends, permeates his letters. He also expressed to her his feelings of disillusionment with the war and its senseless destruction. In turn, Annamarie intimated her experiences as a woman trying to contribute to the war effort. She worked as a nurse and an advocate for disabled veterans and their families – these experiences gave her a sense of self-worth and confidence. Though at one point he did write that he thought these changes for women should only be temporary, Böhme expressed great pride in her intelligence, resourcefulness and hard work for the war effort.

The final memoir in this volume, Emil Schaefer's *From Infantryman to Flier*, is an example of wartime publications that adhered to a more conventional formula. Published by the same August Scherl that competed with other leading publishers to produce pilot memoirs (including Buddecke's, which is featured in volume one of this series), it's aimed at an audience that expected the mythic image of the fighter pilot. In many ways, Schaefer's collection of diaries, notes and letters that were assembled by his father shortly after his death in 1917 fulfills this goal with its emphasis on heroic sacrifice and descriptions of war as an adventure and a game. At the same time, like other wartime publications that fed the military's propaganda agenda, there's a bit more complexity embedded in the narrative. Despite his claims to having his nerves under control, the psychological strain of combat was clearly taking its toll. Tensions between Schaefer and his loved ones on the home front also jump off the page as he struggles to explain what he's going through, especially after being wounded in the trenches before he joined the air service.

But perhaps what readers will find most interesting in Schaefer's narrative is the remarkably rich descriptions of the daily life of a front-line pilot. As one of the early recruits in von Richthofen's elite *Jasta* 11, Schaefer experienced the early successes of that unit and 'Bloody April' 1917. He describes not only the daily rituals and survival techniques, but also the language used by fledging aviators to convey the new sensations of flight – Schaefer dedicates a whole chapter to analyze fighter pilot terminology. It's extraordinary that all of his experiences as a fighter pilot were encapsulated in a few short months before his death in June 1917 at the age of twenty-five.

Similar to volume one of this series, more background and analysis of these pilots appears in short introductions before each memoir. For further in-depth detail on the types of aircraft, personnel records, promotions and other data, I would highly recommend the excellent series by Lance Bronnenkant, *The Blue Max Airmen* (especially volumes 7 and 11, which examine Schäfer and Böhme respectively).

A quick note on the transcriptions and translations: I tried to remain as true to the original text as possible, but I also made an effort to make the translations idiomatic and accessible to English-language speakers, while retaining the original style of the authors. Some of the colloquialisms used by German pilots can be particularly difficult to translate. These cases are noted in endnotes. Endnotes also provide some of the original language, especially where there is ambiguity, unique aeronautical terms and other details. For precision, most military terms, ranks, and locations are provided in the original German, with translations in brackets.

Finally, I'd like to give my gratitude to editor Jack Herris at Aeronaut Books for his assistance, expertise with images, and kind support in publishing this series of translations. I'd also like to thank my son, Max Coolidge Crouthamel, for his technical expertise, enthusiasm for both technology and humanity, and his willingness to stare at artifacts in museums with me. In the introductions to each individual pilot below, further acknowledgements and thanks are given to various institutions and archives who generously granted permission to use documents and images.

I hope readers will find these sources interesting and useful as they provide another perspective into the mentalities of the first aviators and their experiences in modern war. In the case of Rosenstein, I hope readers will gain insight from his account about the consequences of hatred, as well as the importance of protecting the rights of individuals from different backgrounds who deserve respect as citizens devoted to their nation.

Jason Crouthamel,
Grand Rapids, Michigan, 2022

Lt. Emil Schaefer Albatros D.II D.504/16

Lt. Emil Schaefer Albatros D.II D.1724/16

Lt. Emil Schaefer Albatros D.III D.2062/16, *Jasta* 11

Introduction to Willy Rosenstein, "Autobiographical Note"

Willy Rosenstein's autobiography reflects not only an extraordinary wartime experience, but also an exceptional narrative of 20th century German history. Rosenstein was a German-Jewish pilot in the First World War who flew as Hermann Göring's wingman for a short time. He survived combat only to be attacked by the Nazi regime, and he emigrated to South Africa in 1936, escaping Hitler's racial laws. In South Africa he set up an airfield next to his farm outside Rustenberg where he trained pilots, but he was arrested in 1940 and sent to an internment camp as an 'enemy alien' for about six months. During that time, he typed this memoir about his experiences in 1914-1918 and the postwar years. After he was released, he continued to train pilots including his son, Ernest, who flew with the 185 Squadron RAF. Ernest was killed in a strafing run attacking Nazi positions in Italy on April 2, 1945.

The narrative of Rosenstein's life reflects the extraordinary experiences of a German-Jewish veteran from the Great War who was persecuted by the nation for which he fought and was wounded in 1914-1918. While remarkable, Rosenstein's experiences were also typical of those encountered by many German-Jews who fought for imperial Germany. Over 100,000 Jews served and 10,000 died in First World War, at least as many, proportionate to their numbers, as Germans from Christian backgrounds who served and died in the war.[1] German-Jews, like Christian front soldiers, shared the experience of 'comradeship,' which included a sense of loyalty, patriotism and sacrifice that bonded German soldiers. The ideal of 'comradeship' was a crucial component of the masculine ideal, and for German-Jewish soldiers, it reinforced their hope that they would be accepted by the nation they fought for, and that their loyalty would be cemented by their military service.[2] German-Jewish soldiers held on to this faith in comradeship even as anti-Semitism grew during the war. In 1916, the Kaiser and his generals, seeking a way to deflect from their own failure to win the war, ordered the infamous "*Judenzählung*" ("Jew count"). Based on anti-Semitic prejudice that Jews were not pulling their weight in the war, the military counted the number of Jews fighting at the front, with the assumption that they would 'discover' that Jews were not fighting in adequate numbers. The "Jew count" was based on flawed methods. The reality was that Jews, hoping to prove their loyalty and patriotism, served in the same numbers as their

Above: Cover of Theilhaber's *Jüdische Flieger im Weltkrieg* (1924). The cover image depicts the German-Jewish pilot Fritz Beckhardt and his Siemens-Schuckert D.III fighter, which is painted with a white *Hakenkreuz* (swastika), a symbol of good luck used by many pilots before it was used as a symbol of the Nazi party after 1918.

Christian comrades. The results of the count were suppressed by the imperial government, which continued to make false accusations that were based on hatred and prejudice rather than evidence.[3]

In the immediate aftermath of the war, right-wing circles fueled the 'stab-in-the-back' legend, which falsely accused Jews, socialists and other groups long-perceived as

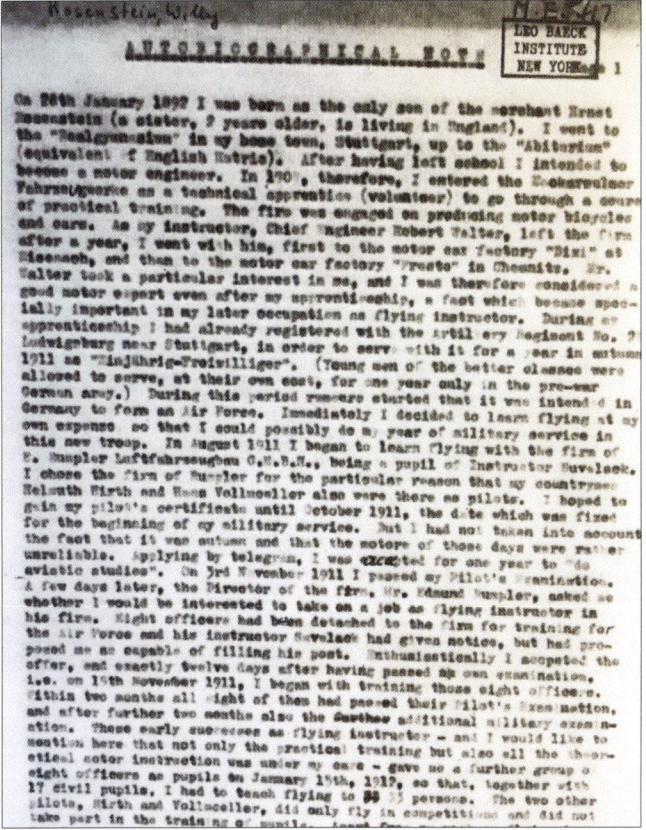

Above: Willy Rosenstein's typed memoir, in the Leo Baeck Institute archive, New York.

Above: Willy Rosenstein photographed as a young boy in Stuttgart. Original in the possession of the University of Cape Town, South Africa, Cape Town Holocaust and Genocide Centre.

'outsiders,' for betraying the nation and causing defeat. Jewish veterans' groups organized to counter these accusations by publishing accounts of Jewish soldiers' sacrifice and heroism. This included the release of the volume by Dr. Felix Theilhaber, *Jüdische Flieger im Weltkrieg* (*Jewish Pilots in the World War*, 1924).[4] German-Jewish veterans thus hoped to prove their Jewish identity was, like Christianity, a private religious identity. Being 'German' through loyalty and sacrifice in the war seemed self-evident. However, shortly after they came to power in 1933, the Nazis attempted to erase the sacrifices of German-Jews in 1914-1918 from the official memory of the war. The Nazis saw Jews as a separate and 'inferior' race, unchangeably 'outsiders' and a threat to the racial health of the nation. The Nuremberg Laws, which culminated in 1935, defined German-Jews like Willy Rosenstein as non-citizens, and pushed them out of the economy, education, medicine and other sectors of German life. German-Jewish veterans fought back against these racist policies. Holding their iron crosses and wound badges earned in the war as proof of their heroism and sacrifice, Jewish men claimed that their war experiences demonstrated their devotion to the nation. Many felt betrayed when they were attacked by the same nation that they defended in the Great War. Nevertheless, the Nazi regime persecuted Jewish veterans along with the rest of the Jewish community, and many were murdered in the mass shootings and the extermination camps during the Holocaust.[5]

As the steps towards the Nazis' 'final solution' unfolded, German-Jewish veterans like Willy Rosenstein struggled to survive. The individual bonds of comradeship that he established with old war buddies helped him to build the foundations for his escape. Rosenstein's experiences reveal the irrationality of anti-Semitic prejudices, as his old gentile comrades, some of whom were staunch anti-Semites and leaders in the Nazi regime, rationalized that *their* Jewish comrades were "exceptions." One can also find different layers of rationalization, apathy, indifference, prejudice, and, on some occasions, efforts by 'good' neighbors who were part of the complex community that Rosenstein escaped from just before the outbreak of World War II.

These different layers of life in Nazi Germany should

Above: Postcard of Willy Rosenstein by S. Fischer.

interest students and scholars of this period. Perhaps most interesting is Rosenstein's relationship with Hermann Göring (spelled "Goering" in this English-written memoir), who became the *Reichsmarschall*, head of the Gestapo and one of the leading perpetrators of the Holocaust. Göring was Rosenstein's commander in *Jasta* 27 in late 1917, and Rosenstein often flew as Göring's wingman. Rosenstein recounts here how Göring's anti-Semitic comments in front of mates in the squadron mess led him to request a transfer. Shortly after that incident, Göring provided Rosenstein a recommendation, which stated: "[Rosenstein] has won the confidence of his squadron commander due to his aggressiveness in air combat and the affection of his squadron-mates because of his fine comradeship."[6] The contradiction between Göring's anti-Semitism and his praise of Rosenstein's "fine comradeship" give us a glimpse into the tightrope that German-Jewish veterans had to walk, and their confusion over whether or not they were accepted as comrades or rejected as permanent outsiders.

When Rosenstein eventually ended up in *Jasta* 40 in the summer of 1918, he felt much more sincerely accepted by his new comrades and especially the squadron commander, Carl Degelow. Degelow's admiration for Rosenstein can be found in Degelow's postwar memoir, in which he described Rosenstein as a cornerstone of the squadron, whom he frequently chose as his wingman for dangerous missions. Degelow wrote: "[Rosenstein] was our 'patron saint,' performing for us the essential duty of keeping our squadron formation free of enemies, as he flew last and highest in our formation."[7] Degelow wrote a letter to his superiors recommending Rosenstein for a Knight's Cross of the Royal Order of Hohenzollern with Swords medal for bravery (Rosenstein ended the war having shot down nine enemy aircraft). Rosenstein and Degelow preserved a lifelong friendship. They met in Berlin occasionally in the 1920s with other old comrades from *Jasta* 40. This circle of friends was crucial to Rosenstein, whose life was otherwise disastrous. He recounts in his memoir the death of his first wife, Hede, and his descent into deep depression raising his young son Ernest, alone. He turned to morphine addiction, sparked while recovering from a wound in 1916. When he lost his business and livelihood in the wake of the Nuremberg Laws, his morphine addiction, which he referred to bitterly as "the cure for everything, M," intensified along with his despair that the fatherland for which he fought had now turned against him.

Above: Postcard of Willy Rosenstein in a Hansa Taube.

While in the depths of despair, Rosenstein was saved by his friendships with his old comrades. He was proud of his old commander, Degelow, who "never concealed his anti-Nazi attitude." But most fortuitous was his chance meeting in a Berlin café, shortly after he'd been pushed out of his job by Nazi racial laws, with his old comrade Friedrich Wilhelm Siebel. Rosenstein knew Siebel from his two-seater squadron before their fighter pilot days, and Siebel was also a friend of Göring. Siebel offered to contact Göring on Rosenstein's behalf, and in turn Göring wrote a letter that helped cut through bureaucratic red tape to expedite Rosenstein's emigration to South Africa. Rosenstein never contacted Göring directly – indeed he had not seen him since the anti-Semitic incident in *Jasta* 27's mess in December 1917 -- but Rosenstein wrote in his memoir: "I must admit that Göring's letter made things easier in some ways." Göring's precise sentiments about his old squadron mate are not known, but this bizarre paradox of a leading Nazi official who spearheaded attacks on Jews but at the same time helped an individual behind the scenes highlights the irrationality of Nazi racial thinking.

Rosenstein's memoir is strikingly different from the other narratives published in this series on German pilots. The circumstances in which he wrote it are harrowing. Shortly after World War II broke out, Rosenstein, a former German citizen, was arrested by South African police as an 'enemy alien.' German-Jewish emigres who had reached England, Australia, and South Africa just before the war were interned in camps (over 30,000 were imprisoned in Britain), as German-Jews were ironically treated as potential German sympathizers or spies during the war. Desperate and frightened, with his son home alone and his second wife having recently abandoned him, Rosenstein typed his "Autobiographical Note" in the internment camp in 1940. In his plea for freedom, he tried to convince his readers that he was anti-Nazi and a productive member of society. This autobiography was never published, but rather it served as a kind of legal document for Rosenstein as he fought for his release. Indeed, once he was released in December 1940, Rosenstein contributed to the Allied war effort by training his son Ernest, who would go on to serve in the South African Air Force and the RAF. Ernest's death in the last days of the war flying a Spitfire against the Nazis was another traumatic blow. Rosenstein died in a flying accident with a student pilot near his farm in Rustenberg on May 23, 1949.

Rosenstein's 19-page autobiography, which can be found in the archive of the Leo Baeck Institute of New York (under ME 527.MM64), is quite unusual. His account paints an image of wartime and postwar experiences not found in narratives published for propagandistic consumption. With its frank discussion of the brutality of the front experience, postwar political and economic disaster, his ongoing drug addiction, and the particular terror faced by a German-Jew living under Nazi Germany, it stands as a rather unique document. Though on one hand it is a very 'matter-of-fact', straightforward narrative, his emotions (especially anger at being forced into an internment camp) and sense of irony about the capriciousness of fate and history bleed through his typewriter. His account should grip aero enthusiasts who are interested in the instrumental role that Rosenstein played in pre-1914 German aviation history, as he trained men who would become Germany's first fighter pilots (one of whom helped fly Rosenstein out of Germany in 1936!). His experiences in two-seater squadrons and then as a fighter pilot are also highlights of his narrative.

At the same time, beyond its usefulness as a source for aviation historians, Rosenstein's memoir provides a unique and candid glimpse into the difficulties these men had in the interwar years. Like other pilots, Rosenstein suffered from shattered "nerves," and his attempts to self-medicate with morphine (as well as by living on the edge as a race car driver who dabbled in Mercedes retail therapy) reveal hidden problems that afflicted many veterans struggling to recover after their combat experiences.[8] The death of his first wife, and lingering pain from physical and psychological wounds, continued to haunt him so that by the time the Nazis came to power in 1933, Rosenstein was a shell of his former self. The focus of this autobiography, Rosenstein's struggle to survive under the Nazis, should be gripping for anyone interested in life under Nazi Germany and the traumatic experiences of Jews who faced persecution and struggled to find another home in a world where anti-Semitism was not unique to Germany. Rosenstein's stress, and problems with drug addiction, did not end when he finally escaped to South Africa, and he recounts the sense of isolation that afflicted him just before his imprisonment in the internment camp.

I've provided numerous endnotes in the text to help provide readers with background and suggestions for further readings in German history to help illuminate the context for Rosenstein's experiences, especially with the rise of the Nazis. I want to make a quick comment about language: Rosenstein wrote the manuscript in English and though his language skills are excellent, the text contains numerous grammatical errors (missing commas, errors in capitalization, minor stylistic errors and inconsistencies, etc). I marked some with [sic] or made minor adjustments in brackets to help smooth-out some of the language for readability. Endnotes are also provided for some of the German phrases or odd word choices that he uses. However, the transcript is otherwise exactly as he typed it. Rather than act as a copy editor, I decided to maintain the manuscript's integrity as a primary source that students and scholars might want to study.

Finally, I'd like to give my gratitude to the generosity of the Leo Baeck Institute of New York for giving me permission to publish Rosenstein's autobiography in full. I'd also like to thank Dmitri Abrahams and Michal Singer at the Cape Town Holocaust and Genocide Centre (University of Cape Town, South Africa) for their generous permission to use several images from their outstanding collection. Thanks also to Robin Smith at the Dayton Holocaust Resource Center for permission to use photos from their website.

I hope that readers will find Rosenstein's testimony to be intriguing from a variety of perspectives.

Jason Crouthamel

Endnotes

1 On the experiences of German-Jewish soldiers in the Great War, see the introduction to Jason Crouthamel, Michael Geheran, Tim Grady and Julia Barbara Köhne, eds., *Beyond Inclusion and Exclusion: Jewish Experiences of the First World War in Central Europe* (New York: Berghahn Books, 2018), 1–17.

2 For an excellent overview of the experiences of Jews and military service, see *Jews and the Military: A History*. Princeton, NJ: Princeton University Press, 2013.

3 On the 1916 "Jew count" or "Jewish census", see Michael Geheran, "Judenzählung (Jewish Census)," in *1914–1918 Online: International Encyclopedia of the First World War*, edited by Ute Daniel et al. Berlin: Freie Universität Berlin, 2015. DOI: 10.15463/ie1418.10684. https://encyclopedia.1914-1918-online.net/article/judenzahlung_jewish_census).

4 The complete citation for this source is: Felix Theilhaber, *Jüdische Flieger im Weltkrieg* (Berlin: Verlag der Reichsbund jüdischer Frontsoldaten, 1924). A translation of this is available as *Jewish Flyers in the World War* by Felix Theilhaber, edited by Elinor Makevet and Dr. Dieter Gröschel, published by Cross and Cockade International, 2019.

5 On the experiences of German-Jewish 1914-1918 veterans in Nazi Germany, see Michael Geheran, *Comrades Betrayed: Jewish World War I Veterans Under Hitler* (Ithaca:

Cornell University Press, 2020). For a riveting study of one German-Jewish family, including the veteran Max Schohl, struggling to survive, see David Clay Large, *And the World Closed its Doors: The Story of One Family Abandoned to the Holocaust* (New York: Basic Books, 2003).

6 The text of this letter from Göring is in Robert Gill's essay, "The Albums of Willy Rosenstein: Aviation Pioneer – Jasta Ace," in *Cross & Cockade Journal*, volume 25, number 4, Winter 1984, 311.

7 Carl Degelow, *Mit dem weissen Hirsch durch Dick und Dünn: Erlebnisse und Betrachtungen eines Kampffliegers* (Altona-Ottensen, 1920), 38.

8 The difficulties with addiction faced by many veterans is still relatively unexplored. An exception would be Thomas Childers' brilliant study of American soldiers returning home after World War II. See his book *Soldier from the War Returning: The Greatest Generation's Troubled Homecoming from World War II* (New York: Mariner Books, 2010).

Above: Lt. Franz von Scheel's Albatros D.II 504/16. Emil Schaefer took over this aircraft after he crashed his first Albatros D.II, 511/16.

Right: Böhme in front of Fokker D.I 185/16 of *Jasta* 2.

"Autobiographical Note" by Willy Rosenstein

On 28th January 1892 I was born as the only son of the merchant Ernst Rosenstein (a sister, 2 years older, is living in England). I went to the "Realgymnasium" in my home town, Stuttgart, up to the "Abituren" (equivalent of English Matric).[1] After having left school I intended to become a motor engineer. In 1909, therefore, I entered the Neckarsulmer Fahrzeugwerke[2] as a technical apprentice (volunteer) to go through a course of practical training. The firm was engaged on [sic] producing motor bicycles and cars. As my instructor, Chief Engineer Robert Walter, left the firm after a year, I went with him, first to the motor car factory "Dixi" at Eisenach, and then to the motor car factory "Presto" in Chemnitz. Mr. Walter took a particular interest in me, and I was therefore considered a good motor expert even after my apprenticeship, a fact which became [e]specially important in my later occupation as flying instructor. During my apprenticeship I had already registered with the Artillery Regiment No. 29, Ludwigsburg near Stuttgart, in order to serve with it for a year in autumn 1911 as "Einjährig-Freiwilliger." (Young men of the better classes were allowed to serve, at their own cost, for one year only in the pre-war German army.)[3] During this period rumors started that it was intended in Germany to form an Air Force. Immediately I decided to learn flying at my own expense so that I could possibly do my year of military service in this new troop. In August 1911 I began to learn flying with the firm of E. Rumpler Luftfahrzeugbau G.M.B.H.,[4] being a pupil of Instructor Suvelack. I chose the firm of Rumpler for the particular reason that my countrymen Helmuth Hirth and Hans Vollmoeller[5] also were there as pilots. I hoped to gain my pilot's certificate until [sic] October 1911, the date which was fixed for the beginning of my military service. But I had not taken into account the fact that it was autumn and that the motors of those days were rather unreliable. Applying by telegram, I was accepted for one year to "do aviatic [sic] studies." On 3rd November 1911[,] I passed my Pilot's Examination. A few days later, the Director of the firm, Mr. Edmund Rumpler, asked me whether I would be interested to take on a job as flying instructor in his firm. Eight officers had been detached to the firm for training for the Air Force and his instructor Suvelack had given notice, but had proposed me as capable of filling his post. Enthusiastically I accepted the offer, and exactly twelve days after having passed my own examination, i.e. on 15th November 1911, I began with training those eight officers. Within two months all eight of them had passed their Pilot's Examinations, and after [a] further two months also the additional military examination. These early successes as flying instructor – and I would like to mention here that not only the practical training but also all the theoretical motor instruction was under my care – gave us a further group of eight officers as pupils on January 15th, 1912, so that, together with 17 civil pupils, I had to teach flying to 33 persons. The two other pilots, Hirth and Vollmoeller, did only fly in competitions and did not take part in the training of pupils. Apart from my work as flying instructor I also had to break in new military machines ("Rumplertauben"),[6] to make test flights, and to fly these machines to the military aerodrome and Döberitz. I also took part in some flying competitions, more or less successfully, which took place on our Johannisthal aerodrome.[7] Until Spring 1913[,] I remained with the firm of Rumpler.

I forgot in the beginning to mention a fact of great importance, which was decisive for my future career. When I was 14 years of age, my father died all of a sudden. Sport had always been his chief concern in my training, something which in these days unfortunately was very often neglected in the training of Jewish children. When I was six, my father brought me a proper bicycle with pneumatic tyres from America, where his business had a branch. The funny thing was that I had to have a bicycle registration card which had to be signed by my father, because I knew quite well how to ride a bike but could not write my name yet. At the age of ten I was allowed to learn riding, and therefore was used to sports when still quite young. When I was 15 I bought a second-hand N.S.U.-Nedel[8] motor-bike from a friend for not more and not less than 40 Marks, but it was only so cheap because it did not work properly. Day and night I fiddled about with it, using textbooks, until greatly to the regret of my dear mother, it worked perfectly (up to 18 miles per hour!!). This motor-bike also made a good runner of me because, whenever the gradient was more than 5 percent, I had to push it up. However, I owe all the foundation to my knowledge of motors – which, through all my life, helped me a lot – to the good old Nedel motor which I took to pieces and put together again at least fifty times. It made me all the more interested in the combustion motor which was then still in its initial stages. When, in 1912, I flew for the first time from Johannisthal to the Mueggelsee,[9] together with Kauders, sports editor of the "B.Z. am Mittag",[10] covering a distance of over six miles (!), I had to think of my old Nedel which just as faithfully as my present 70 h.p. Mercedes motor in the air, had taken me over the distance of 12 miles from Stuttgart to Boeblingen on the ground. (This flight, incidentally, was celebrated in the Press as a cross country flight!!!)[11]– I must apologize for having deviated from my subject. But those details may not be unimportant for the understanding of later events.

In spring 1913 the Gotha waggon factory[12] took up the construction of aeroplanes. I was offered the position as Chief Pilot which I accepted the more readily, as I was particularly interested in testing new types. Such pilots are today called Test Pilots. Today, in the age of aerobatics, such work is probably not much of an enjoyment. But there were no parachutes in those days, and one never knew whether the old thing would fly or not. The planes I had to test all flew – otherwise I would no longer be alive (and not be interned here!). The Gotha Waggon Factory had taken over the existing flying school at Hamburg. This school I had to reorganize, as the lead instructor – for politeness' sake let me call him Mr. K. – would jump from one end of the aerodrome to the other with his "Hansa-Tauben,"[13] but could not fly. I proved in the first test that those machine could fly and that made all the difference. While at Hamburg, I took part in the "Round Mecklenburg Flight" (my first bigger cross-country competition), and was so enormously lucky to win this race by a large margin, at the same time getting all the prizes. In this race I flew one of the first "Gotha-Tauben". After working four months at Hamburg I was recalled to the main factory at Gotha. There I was employed on [sic] manifold tasks. While with this firm, I tested about 14 new types and trained more than 70 pupils. There were four other pilots besides me.

War broke out while I was still with this firm. I immediately registered as a Volunteer Pilot, but to my great regret had to remain with the Gotha Waggon Factory as pilot for another short period. After repeated representations I was transferred to the Pilot Reserve Department (Flieger-ersatzabteilung) Hannover. From there I went to the front, joining the Front Pilot Department (Feldfliegerabteilung) 19, whose commander, Captain v[on] Poser, had been one of my pupils. There is no doubt that I spent a fine time there, especially as the observer in my machine was an old school pal of mine, Lt. Martin. (His father, General, v[on] Martin, fell during the war.) For sixteen months I flew with Martin, until a French fighter pilot shot us down during the Verdun offensive (Nieuport).[14] This happened about 10 miles behind enemy lines. Martin's one upper t[h]igh was shot off,[15] and I was hit in both legs. Most fortunately for us both, I was in a position to control the machine and to land it on our own aerodrome before fainting. I stayed in the field hospital for four weeks, and then got a short home leave, in order to recuperate completely. Lt. Martin, my observer, remained confined to bed for a long time: and as I did not want to fly together with anybody else, I went back to the front as fighter pilot. At first I was commandeered to the Army Fokker Squadron No. 5 in the Champagne [region].[16] After a short time, however I was transferred to the newly-formed Fighter Squadron No. 27 Its leader was First Lieutenant Göring, after the original leader had been shot down the same day I had registered with the squadron.[17] I remained with this squadron from the autumn of 1916 to the end of 1917, nearly all of the time in Flanders.[18] Toward the end of 1917 my nerves began to give in. As my military papers said, "This is easily understood, considering that Lt. Rosenstein has been employed as a pilot for more than six years, and has spent most of the war as pilot at the front." – Moreover, I had had a personal quarrel with Göring, caused by an antisemitic remark in front of all comrades in the Officers' Mess at Isaghem, Flanders. I had been compelled to demand its revocation. These circumstances caused me to apply for my transfer to a Home Formation which was granted after a short time. In December 1917 I was transferred to the Single Seater Fighter Squadron in Mannheim and after two months to the Single Seater Fighter Squadron in Karlsruhe.[19] While with these squadrons, I had sufficient opportunity to recuperate my nerves. In summer 1918 I was fit to return to the front and applied for transfer to a fighter squadron. At this time I heard of a former comrade who had become leader of the Fighter Squadron No. 40 and I was promptly transferred to this squadron with station at Lomme near Lille.[20] On the first flight together with my Squadron Leader Lt. Dilthey, the latter was shot down by our own anti-aircraft guns. The squadron's new leader, Lt. Degelow, remained in his position up to the end of the war.

When [the] war was over Degelow became director of a concrete factory in Pommerania.[21] He maintained contact with me and the other squadron members who were now living in various parts of the country. Every few years we met in Berlin to spend a few hours among old friends. Degelow always remained the centre of the circle, the more so as he had become the leader of the "Stahlhelm" in Pommerania, and quite publicly never concealed his anti-Nazi attitude.[22] The result was that, when Hitler came into power, he was put into prison. Fortunately, before anything happened to him, he was released, probably owing to the fact that he owned the "Pour le Mérite" (the highest distinction in the German army), and that, in those days, nobody dared to do harm to

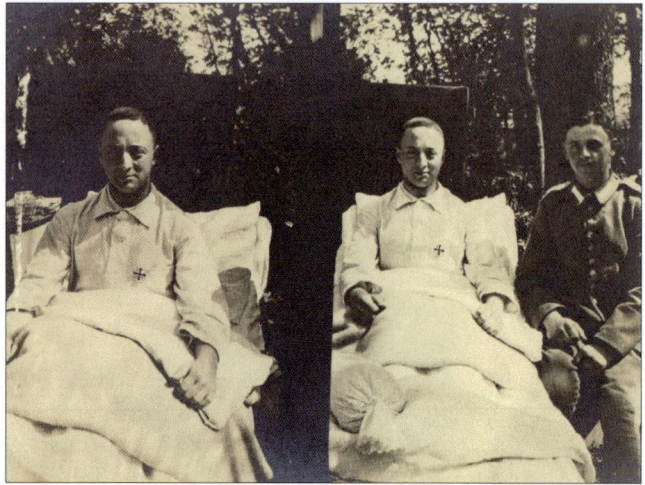

Above: Photo of Rosenstein convalescing after being wounded in 1916. Original in the possession of the University of Cape Town, South Africa, Cape Town Holocaust and Genocide Centre.

such a personality! – after the end of the war I became a member of a corps that successfully took part in crushing the Communist risings ("Spartakists"), as it consisted of old soldiers.[23] And this concluded my military career.

After the war there was no civil aviation in Germany,[24] and I therefore had to look for a new occupation. At first I lived with my parents in Stuttgart. (My mother had become married again. My stepfather was Regierungsrat[25] Karl Noerdlinger, LL.D., who was also President of the "Israelitische Oberkirchenbehörde[26] (the highest Jewish authority) of Württemberg, and I was on excellent terms with him.)[27] When I suggested to learn something about business – of which I had not the slightest idea – he got me into the firm of I. Sigle, Shoe Manufacturers, Ltd. (the well known makers of "Salamander" shoes), as a volunteer.[28] (My stepfather was juridicial advisor to this firm, and was a personal friend of the business director, Consul Max Levi, and the technical director, Mr. Jakob Sigle, P.C.) I entered this firm early in January 1919 and, in the course of one year, worked in all business and technical departments, so that at the end of the first, so-called apprenticeship year the head of the firm offered me a position as his secretary. I need hardly explain that I accepted this offer with enthusiasm, as, apart from being very well paid, I had the chance of gathering further experience under his guidance (he was truly a "Royal Merchant").

An incident occurred in those days which should, as already on a previous occasion, be of the greatest influence for my future life. When invited for dinner by some Stuttgart friends my neighbor at the table was Miss Hede Rothschild, the niece of my employer. In the course of conversation we both found out that we were very keen Tennis players, but were suffering from the lack of a great partner. We agreed to arrange for a game, but I must confess that I was very skeptical as to this experiment, because I had experienced that ladies often looked at a game of Tennis merely as an opportunity for a flirt. But I was to be thoroughly wrong this time. Miss Rothschild, though delicate and small, beat me by a wide margin, though I was a good average player. For several weeks we played regularly, until one day our game found an aprupt [sic] end, when, for reasons unknown to me, Miss Rothschild's mother did no longer permit her to play. I must confess that this was a hard blow for me, as I had terribly fallen in love with her during these weeks. She told me later that she had experienced the same. For the better understanding of what is to follow I think I had better explain the reasons which led to the end of our games – they only became known to me later. As a former pilot I had – I do not want to say whether rightly or wrongly – the ill reputation of being a "Don Juan", and the so-called better-class mothers were doubtful as to my qualifications to become a good husband. This theory had, of course, also come to the knowledge of Miss Rothschild's mother, and the latter (my future mother-in-law) had used all means to prevent our playing together. But one should never try to play with Fortune: things often take a different course from the intended one…

At any rate we were now separated for an indefinite time, and whether by chance or purpose, during the next two years we only met a few times and talked to each other still more rarely. This was not a very pleasant time for Miss Rothschild's mother, because her daughter had seen through her tricks, though she never said anything to us about it. I thoroughly dug myself into my work. Perhaps I owed my quick advancement in the firm during these two years to the fact that my unhappy love kept me away from all distractions, though other little incidents, which I shall mention later, also played their part. My dear mother, as I should find out later, was the only one who looked through me and who knew why her son, who normally likes to go out, suddenly wanted nothing else but to stay at home and to play cards and other games with his father and mother. I imagined that she believed me when, being asked, I told her that I had got sick and tired of going out so much. The next 1½ year was full with work, and there is not much to report about it, perhaps with two exceptions. Firstly, I was allowed several times to go to Vienna with the head of the firm to take part in exceptionally interesting great conferences, which, in my future business life, were of great use to us because it was here that I learnt not to be bluffed by the calmness and apparent uninterestedness of one's opponents. Here I would like to relate one of the abovementioned instances as it was not without influence on the course of events. My office in the factory adjoined to that of the director because he always wanted to have me close at hand. Above us there was a large machine hall. On the one hand, this was rather disturbing (especially in the beginning): as time goes on one gets used to anything…except being interned!!!!), but on the other hand it had the advantage of letting us know at once if something should go wrong with the machines. I must not forget to mention that the Salamander factory had its own power station for driving the machines, which was situated in a central position of the factory. One day I suddenly heard the machines above us going slower and slower. There must be something wrong in the power station. I ran there and found the first machinist, who really should have nothing to do with the boilers, firing away for all he was worth, thus doing the work of four stokers. On being questioned, he explained that these stokers had suddenly refused to work and had run away because they were not satisfied with their wages. We could, so he said, of course not maintain the pressure at the necessary 180 lbs. It had already dropped to 135 lbs, and would rapidly go down more if he did not get immediate help. As no time had to be lost, I took off my coat, turned up my white shirtsleeves, and helped him. I only took this for granted, but my director, as you will see just now, was of a different opinion. Before long, the omnipotent person personally hurried to the power station to blow he stokers up, and rightly so, for their bad work. However, he did not get far with his admonition when he saw what was going on. Our machinist, just as his building, was an exceedingly clean man. But what did he look like! And at his side there was a second "mulatto"[29] whom at first he did not recognize as his secretary. When I looked into a mirror afterwards I understood the fright he got. Within a few minutes some assistance had come. Meanwhile we had driven up the pressure to 165 lbs., so that even a temporary standstill of the factory, [in] which in these days we employed as many as 8000 workman, had been avoided. Apart from this, the enigines [sic] would have been terribly overcharged if any damage had occurred. As I mentioned above, I took my behavior for granted, but my director considered it such an exceptional deed that he asked my father to come the same day, in order to tell him about my "heroics." I can only explain this with the fact that he was a "Mann des Geistes" (an intellectual), and an exceptional one at that, who considered it as something very particular, if an employee who was intended only for office work suddenly did the most exerting manual labour. (As a special reward he had me taken home in his private car after I had a quick wash, and I also was allowed to have the rest of the day off.) But the real reward only came a short while later!!! It would go too far if

Above: Photo of Rosenstein on his Albatros D.III, Jasta 27, in 1917, from the Dayton Holocaust Resource Center.

I were to tell you other little incidents in this short sketch of my life, though they too have played their part in the future development of my life. I remember them whenever I go through my life in my mind, and perhaps one day I shall have leisure to write down all these episodes, if not for publication, then for my son and for a small group of friends whom I hope to have in future days. Again I must apologize for a small digression. The above incident took place toward the end of February 1921.

One Saturday morning, in the middle of March, my director called me into his room and offered me a chair [,] something that, as a rule, only happened for larger discussions. He offered me a cigarette, a fact which was still more surprising, as I never smoked [while] speaking on business matters with him. What has happened now, I thought. Suddenly, a hideous thought came into my mind. A few days ago I had unbosomed[30] myself to my confidant in the firm, the oldest clerk ("Prokurist"), Mr. Hammer, of course without giving any names. I told him that, much as I liked my work in the firm, I would not be able to remain with it much longer, things even my parents did not know about my decision. I asked him not to talk about the matter to anybody. Should my good old friend have gossiped? Later I have made an inward apology for having had doubts in his

Above: Photo of Rosenstein presented with victory laurels by an unidentified woman at the Karlsruhe airfield, June 1918, from the Dayton Holocaust Resource Center.

honesty for a second. To cut a long story short, my director asked me quite bluntly whether I was engaged: he said he had been told so. When I declared this to be nonsense, he invited me for dinner and asked me, with a twinkle in his eyes, whether I would mind to meet Miss Hede Rothschild there. My reply cannot have left him guessing. This was the end of the interview and I went straight into the office of my friend Hammer, because I felt that, having suspected him so injuriously, I must talk a few words with him. Hammer looked at me and asked whether I had won the Irish Sweep (Grosse Los),[31] because that was what I looked like. I gave my reply in the form of the Pythian Oracle:[32] Not yet, but almost – this I left the puzzled Hammer, who only heard the solution [to] the riddle on the following Monday.

To describe the events of this Saturday, so significant for my future life, would necessitate an exposé in itself, and I must therefore try to be as concise as possible. I think a telegraphic style is the best in this case. Dinner at the Hotel Marquardt, surrounded by inquisitive friends, shaking their heads at the strange juxtaposition at our table. Pleasant conversation, coffee in the lounge, still more hand-shaking on the part of the friends, amusing return home: first taking home Director and his wife, then taking home Miss R., under very conventional talk, declining offer by the driver to take me home ("I must have some fresh air"). On my way home I vainly try to find an explanation for this sudden change. I slept very little that night, but made plans for my possible future and a programme for the following day.

Here the result? Got up early, went to town, bought flowers, drove to the house of Miss R.'s parents, met parents and herself at home because really much too early for a conventional visit, for politeness' sake <u>one</u> minute of conversation, mentioning the pleasant evening yesterday, then taken [sic] deep breath… asked Father Rothschild for his daughter's hand in marriage (without ever having talked a single word about it with HER). Her parents ask HER what SHE thinks of it, SHE says "yes," parents say "yes". Happy end!!! We are engaged. – Hede, as I shall call her from now onwards, goes to the phone with me and we inform my parents of the happy event. This was what one used to call a "Husarenstreich." For us it was a "Fliegerstreich."[33]

We two became the happiest couple.

This happened in the middle of March 1921. On 8th May, 1921, that is, about seven weeks later, we were married. Our four weeks' honeymoon we spent at the Insel Hotel at Lake Constance, and <u>not</u>, as usual, in Italy. After our return I entered my father-in-law's firm, the Leather Factory Zuffenhausen, as partner. For reasons of health he had retired from active partnership sometime before.

While with the Salamander factory, I had become a passable merchant and leather expert, but I knew next to nothing about the leather manufacture in a leather factory. Therefore, before beginning my work at the leather factory Zuffhausen, I went to Furth-im-Walde for two months, into the tannery of Mr. Perlinger, in which my former director and now uncle Levi was interested. My wife, of course, accompanied me and, with admirable courage, endured the unpleasantnesses of a stay in the only local "hotel." The fleas nearly devoured us, and only by putting inches of insect powder into our beds, we were able to sleep at least part of the night. But apart from this plague, our stay there was lovely, and the glorious excursions into the surrounding mountains will for ever [sic] remain in my memory. I had taken a motor-bike with me, and with Hede in the pillion seat,[34] and young Perlinger for company on another bike, we went away every weekend. For instance, we drove to the foot of the Arber (a mountain),[35] left the bikes there, and climbed up during the night. During the week I worked as an ordinary tanner in the factory under the supervision of both Perlingers, particularly Perlinger junior. I did the work of an ordinary tanner, and owe it to both of them if I became quite proficient in the course of two months. My reason for going into another factory in order to learn something about the tanning trade was that I was of [the] opinion that it is entirely wrong to learn something in one's own factory where afterwards one intends giving technical instruction. I have experienced that this tends to undermine the authority. I also think a son should never learn his trade in the firm of his father.

I want to mention here that I would have much preferred to remain with the Salamander factory. But it was my father-in-law's urgent desire that I should take over his post, and I gave in somewhat reluctantly. I remained with the factory from 1921 to 1927, and then left, after a friendly agreement with my partner and uncle Arthur Levi (with whom I did never get on too well) had been concluded.

In those years I had to go through the most difficult part of my life. On 20th January 1923 my dear Hede gave birth to our first and only son. She had to suffer terribly during her pregnancy, as a kidney disease, unfortunately recognized far too late, brought her close to her grave. In the eleventh hour her life was saved through the great skill of two medical men, Dr. Moerig, of Stuttgart, and Professor Voelcker, of Halle. I would not like my deadliest enemy to go through the agonies I had suffer[ed] those days and nights! She owed her life solely to the truly heroic and self-sacrificing effort of Dr. Moerig especially. After the birth her state of health improved rapidly, and the baby, too, developed excellently. It would go too far if I were to give a detailed account of the years that followed. They passed in wonderful harmony. Hede was the ideal wife, living for her husband and child only, and thus making it easy for us to live for her only. Therefore I could face the future with confidence. But things should come [sic] otherwise.

In the spring [of] 1926, Dr. Moerig, in whose continuous treatment Hede had been, became very ill quite suddenly and died after a very short time of Miliary Tuberculosis, a terrible disease.[36] Another kidney specialist was immediately appointed to take over the treatment. But he does not seem to have been quite as thorough as our old doctor, and Hede did no longer feel quite as fit. Towards the end of May of that year a business trip took me to Bavaria, and as usual, Hede was with me in the car, which I always drove myself. On our way back we had got as far as Nuremberg and intended to return to Stuttgart the next day. But during the night at the hotel, Hede had the same kind of attack she had had before the birth of our son. She categorically declined my intention to consult a local doctor or to go back by train and fetch the car later. She insisted on returning with me the next morning, pretending to feel much better. I seated her as comfortably as possible in our well-sprung Lansia[37] car and drove back to Stuttgart at very slow speed. My intention was to take her to the hospital and – as I afterwards did – to stay there with her, but again she insisted on remaining at home because she had her child there. I have afterwards asked myself why Hede wanted no doctor at Nuremburg and why, at first, she did not wish to go

Above: Willy Rosenstein and his Fokker D.VII, *Jasta* 40, summer 1918. From the Dayton Holocaust Resource Center.

to hospital in Stuttgart. Knowing her as I did, I have come to the result that, without letting me know, she must have felt how seriously ill she was! As her condition went from bad to worse, she gave in when I urged her, and thus we moved into hospital so as to have a doctor close at hand. I called Professor Voelcker by telegram from Halle to Stuttgart, and he assured me that her condition, though serious, was by no means hopeless. He stayed with Hede for two days, but unfortunately had to return to Halle, being head of the Surgical Clinic of the University where he had to attend some very urgent cases. But he promised that he would come back to Stuttgart the same night if I called him by telephone, and he kept his promise. Two days and nights passed in suspense, without any change in Hede's condition taking place. She suffered with a heroism, which even today seems to me to have been super-human, and she even was capable of cheering me up. I think this is why I did not regard her condition as hopeless as it really must have been. On 7th June, in the evening, her condition suddenly deteriorated so that even a layman could see it. I immediately phoned to Halle, and Professor Voelcker promised his immediate departure so as to be in Stuttgart the next morning. I went to my dear Hede with this consoling news, because I knew that she thought just as highly as I did of Professor Voelcker and had great confidence in his medical knowledge, which once before had helped her.

The next few hours were the saddest in my life. Helplessly I had to watch Hede become more and more feeble, and though the doctors did all in their power and fought heroically for her dwindling life, Hede died in my arms at almost two in the morning. Up to this moment I had pulled myself together, especially as the doctors had told me how Hede had been fighting up to the last minute and had been nearly fully conscious. But the terrible reaction

Above: Willy Rosenstein (seated, second from right), with his *Jasta* 40 squadron mates in 1918. His comrades are horsing around, pointing guns at the squadron commander, Carl Degelow (seated, second from left). From the Dayton Holocaust Resource Center.

came immediately and I fainted. How I got into my parents' house that night I could not tell. At any rate it was a good thing I was not taken to my own home, because then I would probably not be in a position to write these pages today. A further good measure of my mother's was that on our way home to her house she took my son with us who was then only three years old. When I woke up in the morning he was lying at my side in his cot and laughed toward me with his big eyes. He thus made me understand that I still had a responsibility and was not entirely left alone in the world. Professor Voelcker was to arrive in Stuttgart early in the morning; therefore I pulled myself together and went to the station to break the sad news to him. When he alighted from the train he looked at me with questioning eyes, and up to this day I remember the sad commotion in the face of this great scholar, who was moved to tears by the death of a distantly known human being. He went home with me, and stayed there for the rest of the day, though he could have returned to Halle immediately. He assured me again and again that everything had been done in this case, but that Nature, after all, was stronger than men. A post-mortem which, following Voelcker's advice, was made in the interest of our son, showed that my poor Hede had been suffering from an organic defect which sooner or later must lead up to a catastrophe. This was a small consolation to me as thus she had been spared a long suffering. The day after Voelcker's return to Halle I received a letter which was partially the most moving document I have ever had in my hands. He assured me that never before had he been as unhappy that his medical skill had not been able to save a life, and that he could not prevent our happy marriage from being pulled to pieces, a marriage of rare harmony. He said that he had come to like Hede when he had admired [her] for her courage. Thus, this widely occupied scholar had gained an impression of our life in the few days he had spent with us. He was not only a great doctor but also a great judge of human nature.

In the two days remaining until Hede's funeral I fortunately did not have much time to ponder because I had to fulfill an important task which I would almost like to call a mission. In Furth-in-Walde, where Hede and I had spent my time as apprentice in a tannery, as mentioned above, we had one day visited the local cemetery which, because of the position of the lovely mountains, was one of the few sights of the place. We had been particularly impressed by the graves of some families of standing, not because they were just graves but because, surprisingly for such a small place, they were real tombs. Immediately we had both been taken with the idea not to be put simply into the ground but to be buried almost princely in a proper tomb. We had taken the decision that, should we be able to afford it, we would be buried in such a way. At that time I had not dreamt that only five years later I would have to have such a tomb built in the greatest hurry. After Voelcker's departure I had only two days to bring about my decision. My dear Hede's wish had to be fulfilled at all costs. Scarcely had I taken this decision when a steady downpour began which threatened to destroy all my good plans. But where there is a will there is a way. I stayed on the cemetery almost day and night, and with a lot of friendly talk I managed to keep the masons at their work. Two hours before the funeral the tomb was finished, and a great load had been taken off my heart. Thus this wish of my dear Hede had been fulfilled at the eleventh hour.

How I got over the funeral I do not know to this day: everything [a]round me was like in a mist. But even the most difficult hours in one's life, which seem entirely unendurable, pass by and life puts up new demands. During the day I had my work in the firm, and when I came home there was my boy whom, as was only natural, I loved dearly as an heirloom from Hede. But the nights were terrible. After having tried in vain to find rest, the only way out was to take my car as soon as the boy had gone to sleep and to drive across the country. Dead tired I got home towards morning and slept heavily for one or two hours. At six-thirty sharp I drove to the factory where the lorry driver who cleaned my car would tell me how many miles I had been doing during the night. The distance covered usually was between 150 and 210 miles. Everybody still understood that my nerves, strained as they already were, became badly overworked by such a way of living. It is therefore not surprising that I took more and more drugs, harmless ones to begin with, but more and more strong ones later, until at last I arrived at the utmost remedy, Morphin[e].

In January 1927, i.e. about half a year later, I left the leather factory Zuffenhausen. This was the more easy [choice], because, as I forgot to mention, my father-in-law, my mother-in-law, and my uncle Max Levi (my best advisor) had died during 1924–25. In the same year I suffered another great loss when my dearly-loved mother died of a stroke a few hours before leaving for a holiday at Nice. Thus my last support had been taken away from me.

I took over, in the same year, a small compressor factory, which belonged to a friend and in which later on chiefly spraying plants were produced according to my design. But I was not fully occupied in this new work. On a short trip to Lucerne, together with my step-father and my son, which we made in newly-bought Mercedes S car, I made the acquaintance of the well-known racing motorist Hans Stuck. Following his suggestion I took part in the Kriens-Nigenthal race which took place just then, and, always seeking for [sic] new stimulants, took an immediate fancy to racing. For three years, from 1928–1930, I was a faithful adopt[ee] of this new sport. After the first successes I changed the S car with an SS one, which was much faster (top speed obtained in the flat race near Rastatt: 150 miles per hour). Later on I changed this with an SSK car which was still more adaptable to mountain races.[38] In these three years I got as far as gaining over 25 races in Germany, Austria, Czechoslovakia, and Switzerland, often beating professional racing car drivers. The German press called me the then best gentleman driver of Germany. (Paper cuttings are in my possession and may be inspected.) For the racing season 1931 I would have had to buy a new car. But I had already bought three cars (one of which I had smashed in the Buckow race. Bumping into a tree at 60 miles [per hour][39] without hurting my [sic] or my companion but hitting the tree off ((!))),[40] each of which cost about £1800. I thought I had spent about enough money on this marvellous [sic] sport and reluctantly decided to give it up in order to devote all my time to the factory which had developed satisfactorily. During my absence my faithful manager Jörger had conducted the firm. His name must not remain unmentioned here because up to my emigration he stood faithfully by me. Quite unconditionally he took my side on three occasions in which I came into conflict with Nazi authorities, and in each of them the consequences might have been very serious. More about this later. In 1931 my step-father, with whom I had been on most confidential terms, died within a few days of cancer. Thus I was left quite alone with my son. Apart from this loss, 1931–2 passed rather uneventful, but I looked at the continued growth of the Nazi party with great alarm, as it was connected with an ever-increasing anti-Semitism.[41] My intention to move to

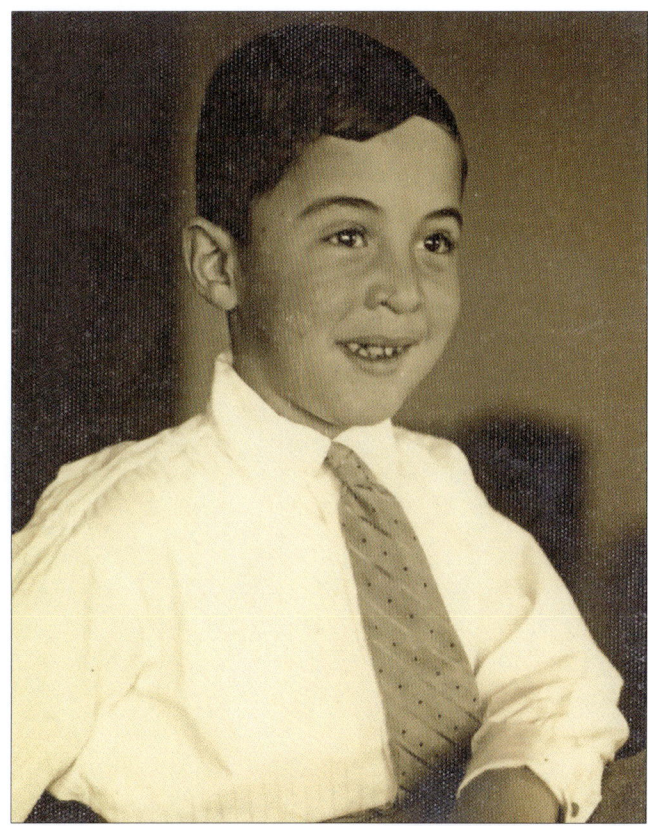

Above: Willy Rosenstein's son, Ernest, in the 1930s. At his airfield in South Africa, Rosenstein trained Ernest to fly, and Ernest joined the RAF during WWII. Ernest was killed flying a Spitfire in a strafing run against Nazi positions in Italy in April 1945. Original in the possession of the University of Cape Town, South Africa, Cape Town Holocaust and Genocide Centre.

Switzerland, which had always been my dream, could not be carried out for the following reasons. The largest part (almost 90%) of our property consisted in Salamander shares out of the estate of my parents-in-law and my uncle Max, who had provided for their not being sold before 1936. His good intentions had been not to endanger the existence of his life's work by a possible premature sale of large amounts shares by one of the members of the family. Even this exceedingly clever man, when he made his Last Will in 1925, could not have foreseen the course of events.

My son Ernest, named like his grandfather (my father), had just passed the elementary school, and was entering the same High School whose pupil I had once been – my favorite teacher having become head master and particularly looking forward to having Ernest in his school – when the catastrophic event of Hitler's advent took place. The state of my nerves was such that I myself felt it could not go on like this. More or less I lived exclusively on drugs. My doctor found a small sanatory at Lindau on Lake Constance

Above: Willy Rosenstein climbing into his Piper Cub on his airfield in South Africa, shortly before his death in a plane crash in 1949. Original in the possession of the University of Cape Town, South Africa, Cape Town Holocaust and Genocide Centre.

whose doctor ensured me over the phone that, if I was not a particularly complicated case, he could heal me within ten days. (Naturally, my doctor had previously informed him in detail of my condition in a letter.) This, I must admit, seemed an incredibly short time, but it seemed worth a trial. Two days after our conversation over the phone I had arrived there by car (it is only four hours from Stuttgart) and had made his personal acquaintance. His name was Dr. Speer, and from the first moment I had unlimited confidence in him. He was a friendly Bavarian and did not look by any means what I had expected a nerve specialist to look like. I stayed at his clinic for the exact period of ten days [as] originally intended, and left it as a healthy man. For a long time afterwards no drugs entered my body and my state of health became better and better. When I left him he had given me the advice to receive treatment for some time from a nerve specialist in Stuttgart, and I did this the more readily as I had an a [sic] good friend the well-known nerve specialist Dr. Hugo Levi. He visited us daily, and by a mere talk with him, usually for about half an hour, my nerves steadied and improved more and more. Then almost suddenly an almost funny reversal took place in our relations[ship]. One day Dr. Levi arrived in a very depressed condition, almost with tears in his eyes, and told me that he would probably not be in a position to treat me much longer, as he could no longer stand the mental strain of living under a Nazi regime, and would therefore emigrate.[42] (Dr. Levi was a man of great means who was not forced to leave Germany for financial reasons, but because he had been excluded from medical societies in spite of long service at the front, etc). He was in such a state that on this day he did not have to console me, but I had to do it with him. Of course I was very unhappy in view of the impending loss of my mental advisor, more particularly as I had suffered a serious gallstone attack only the other night, which he had cured with several hours' work without taking recourse to the otherwise usual medicine Morphin[e]. These gallstone attacks unfortunately took place at regular intervals, until a thorough examination proved that there was a large gallstone.

Shortly afterwards I said a final good-bye to Dr. Levi, as he left Germany together with his family. I would not make up my mind to go to another nerve specialist, and therefore decided to follow Dr. Speer's sound advice which he had given me on my departure from his clinic. His advice had been to do as much sports as possible as soon as my general physical condition, which had of course suffered during the severe cure, permitted it. Again I was lucky to find a young sports instructor, Mr. Schnoldas, a Czech, who most seriously devoted himself to the coaching of the old gentlemen – after all I was over forty [sic]. The training took place in a small swimming bath, and I lasted 1½ hours every day, precisely, the time being spent thus: 15 minutes physical exercises, 15 minutes running with him as pace maker (!!!), ½ hour tennicoits,[43] 15 minutes swimming (the time having to be improved by one second every day: he was there with a stop watch), and finally a massage of fifteen minutes, most strenuous massage, but without hurting me. For these months I stuck to this plan with iron energy. Then suddenly Mr. Schnoldas accepted a position as trainer with a great sports club. Again I had lost a great helper. Meanwhile, however, I felt that I had become so fit that the idea came to me: What about having a try at flying again? My old co-pilot Helmuth Hirth had several private machines in his hanger and would certainly lend me one. He readily agreed to it, and off I went to Boeblingen where his machines stood. And lo! I could still fly! For a few weeks I regularly flew from seven to seven-thirty in the morning, and this gave me the confidence that I was quite fit again. But again the Nazis thwarted me. The man in charge of the aerodrome was a former fellow-officer, First Lieutenant of Police St---.[44] I was on intimate terms with him as he was one of the finest characters I had ever met. But it would go too far if I were to go into more details, and I must simply ask [you] to accept my statement as a fact. To cut a long story short, this comrade one day was told not to have any further dealings with me because I was [a] Jew. As a matter of fact, only an attempt at this was made, because St--- refused to obey this order. Against my will he took the matter to the highest authority in Wuerttemberg, the "Reichstatthalter,"[45] by whom he received a letter of which I would like to quote one sentence. This is how it ended: "Basing on the above, let [sic] Lt. of Police St---- cannot be expected to give up his friendship with the factory owner Mr. Willy Rosenstein which was formed during the war: but he is requested to shape it in such a way, that National Socialist circles will not be offended by it." In the rest of the letter I was always mentioned as "the Jew Rosenstein." In a word, when St--- was wearing [a] uniform he was not to be seen in my company. This mean attempt to destroy our friendship completely spoiled my enjoyment of flying, so that I gave up this beautiful sport reluctantly – only as a sport I had looked upon it. St--- almost got annoyed with me about this, but at least he understood my objection, which was that he as a married man with two children simply had to consider his family. But as often as he came to Stuttgart, he never omitted to call at my house, to prove his loyalty. He was one of the last who shook hands with me before my emigration, and this old soldier had tears in his eyes of which, indeed, he need not be ashamed. Unfortunately, such characters were very rare in the "New Germany."

At about this time I met an acquaintance, Mrs. Blum, at a little dinner at my cousin's. Mrs. Blum, who was married at the age of 17, was divorced from her first husband after a short but very unhappy marriage. She was as clever as she was charming, and during the evening for the first time since the death of Hede, the idea of a second marriage came into my mind. I considered myself in perfect health, and I gave more and more time to these thought[s]. Another circumstance played its part: Mrs. Blum had a charming son, only two years younger than Ernest, who could become an excellent comrade for him. This would have ended my continuous sorrow at seeing Ernest grow up as an only child, and the same would have applied to Mrs. Blum's son. At this opportunity I will not omit to say that during all three years it had been quite clear to me how great a danger it was for a child to grow up alone, somewhat coddled at that. I had to be Ernest's father and mother, and therefore could perhaps not always be strict enough, but during the first years after Hede's death a second marriage appeared to me like sacrilege. Time, however, heals the deepest wounds, and there did not seem to be anything monstrous in the idea now. Once this then got into my head, I could not get rid of it.

Above: Cover of Willy Rosenstein's Flying Log Book, acquired by Jason Crouthamel from an antique bookstore in South Africa. Dmitri Abrams at the Cape Town Holocaust and Genocide Centre (University of Cape Town) thinks that the log book had been somehow separated from Rosenstein's effects after his death and ended up in bookstore rather than Rosenstein's collection at the University of Cape Town.

Our chance meeting turned into one looked-for by me. We became good friends, as did our children; and after a year Mrs. Blum became my wife. (For queerness' sake I would like to mention that Mrs. Blum's husband had also been a leather manufacturer. In Vienna, Austria.) We became married in December 1934.

I forgot to mention two incidents which happened in 1934 and which could easily have been fatal. During the war I had a mechanic, Paul Heinrich. He was living in Upper Silesia and was very badly off. In 1927 I had got him and his family to Stuttgart as my chauffeur. He was a decent fellow, though strongly influenced by his ambitious wife. In the evening [of] 1st May 1934 a friend came to me and told me that he had seen Heinrich and his family taking part in the May Demonstration of the Nazis.[46] I had been a little surprised that he had asked a day off, something that hardly ever happened, because whenever he had any business

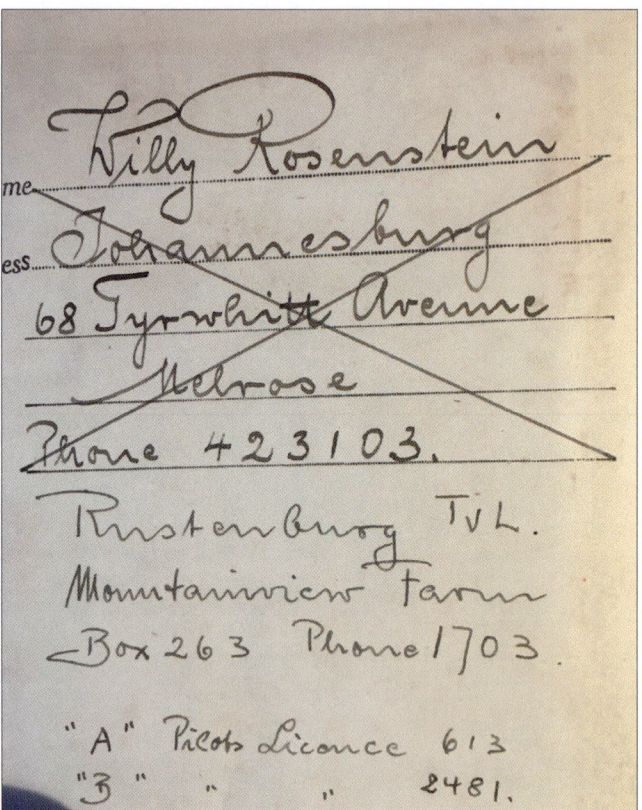

Above: Rosenstein's autograph and address in the front page of his log book.

in town, he or his wife (whom I gave a life in such cases) did it during the week. I saw red when I heard this bit of news[,] though I still thought that it might be the result of some mistake or even calumny. Heinrich was living in my house, and I could therefore easily find out. I called him and told him what monstrosity – such I considered it to be – he had been charged with. To my great consternation he became white as chalk and admitted to have taken part in the demonstration. More than that: as I had come to know it, he also had to admit that he had become [a] member of the Nazi Party some time ago. I left it to him either to leave the Party or me and gave him 24 hours to make his decision. I told him he would understand that I could not have a chauffeur who was a Party member in my house. The next morning he informed me that he would leave the house the same day, giving up his position with immediate effect. Indeed he left the same day, and only a few days later I understood why he was in such a hurry. I received a letter from the Nazi Party in which I was asked to appear on a certain day in order to justify my action of dismissing a servant for the only reason that he was a member of the Party. On the appointed day I took my manager Mr. Jörger with me, because I did not know whether I would safely get home or, as had happened in other cases, might disappear somehow. After a very violent discussion in which I insisted that I as a Jew could not be expected to give employment to a Party member, the position became rather critical, and I said I was prepared to pay Heinrich a certain compensation – which, I am sure, these party officials have shared with him. This was how he showed his gratitude for all I had done for him and his family in all those years. But had I not paid him this sum, I might have fared badly. Mr. Jörger, days afterwards, assured me that during the discussions he felt very sceptical [sic] of my safety, as I had definitely spoken too sharply. Scarcely had the excitement died down a friend, when I had told about this case, informed me that this was not the only Nazi in my service. I laughed at him, upon which he informed me that the brother of my secretary (she had been in my service for many years) was also a Party member, and he belonged to the National Socialist Motoring Corps. His name was Emil K., and he was employed in the office of my factory. Now the stone had again been set rolling, and though Mr. Jörger urged me to dismiss Mr. K. for some other reason, I could not make up my mind. I told him straight that he was a Party member and he had to admit this rather ashamedly. He said his sister who had recommended him to me knew nothing about it, nor did his parents, and they must not know anything because otherwise they would break with him. Anyway, I need not mention that he was sacked. However, he said nothing about it to the Party, but was careless enough to inform a comrade, who had nothing more urgent to do than to consider it his duty of honour to denounce the Jew. It would go too far if I were to describe the abominable consequences. The final effect again was that, in order to avoid serious trouble, I had to pay compensation, thus having a narrow escape. These two experiences which could easily have been fatal, were enough for me, and I therefore became more careful.

Soon after our wedding we saw ourselves compelled to take our boys away from school. The racio-political classes[47] took a form which one could not expect a child's mind to tolerate if the child happened to be Jewish. Let the Nazis poison Aryan children's minds, but not at the expense of our children. With a quick resolve I visited my old master[48] Breitweg and asked his advice. He too agreed that he would not leave children for a minute more at such a school. He was of the reliable old type and shortly afterwards had to give way to a party member. My wife and I immediately went to Switzerland and inspected several schools. When we came to the "Country School Hof Oberkirch" we were immediately taken with the personality of the vice-headmaster, Dr. Wilker, who, as he told us later, had had to leave Germany in 1933 because of the Nazis. He was a member of the Society of Friends.[49] Before we had spoken much with him, we had both made up our minds to send our children to this man's school. They both remained there until our emigration, and spent a marvellous [sic] time, learning much of practical value. All students had to take a carpentry course under the guidance of a professional carpenter who belonged to the staff of the school. A big swimming bath and football ground were taken for granted. We were never sorry of [sic] our choice. The only difficulty in the beginning was that of being separated from our children. This was a particular hardship for me, who had never been without Ernest. Everybody will believe that I did not like the Nazis better for separating me from the children. But the thought to have spared them, many unpleasantnesses [sic] helped us, patiently to bear this handicap. As often as the school regulations permitted (these were reasonably rather strict so as to save children from too frequent farewells to their parents), we paid them a visit by car, as it was only four or five hours from Stuttgart. During the holidays[,] we regularly went into the Swiss mountains with them, which we all loved so passionately. Only in the most urgent cases should the children re-enter Germany, and this, thank God, could be prevented. Thus the end of the year 1935 came near. The chicaneries with which we German Jews were bothered became worse and worse. A new regulation prohibited housemaids of less than 45 years of age from serving with Jews.[50] Our maid, who had been with us for many years but was not quite as old, took leave of us with tears in her eyes. Then my wife declared that she would not dream of ridding the Nazis of their old servants whom they did no longer want.[51] I fully agreed with her, and from that day onwards we did our housework alone. And it worked! We spent Christmas with the children early to bring them back to school afterwards.[52] Conditions became more and more unbearable. Especially my wife, who was more sensitive, suffered under them. (I ask permission to use her pet-name in [the] future. It was "Flower," a partial translation of her former name "Blum,"[53] as she hated her Christian name Paula.) Meanwhile, by an agreement with my relatives, I had been able to obtain control over the Salamander shares, and I might have been able to dispose of them.

Then suddenly, one day, Flower had a proposal to make during supper. "What would you think if we were to make use of the time left till the children's Easter holiday in April for a trip to South America or South Africa[?]" I must have looked at her a little bit puzzled, but she told me that she had already got information about the travelling facilities in the meantime. The last Zeppelin would leave for South America in a few days' time, whereas one could travel to South Africa with the "Boschfontein"[54] of the Holland Africa Line, leaving Rotterdam on 28th January. After a four weeks' stay one could travel back on the "Anchizes"[55] of

Above: Rosenstein's list of various maneuvers in his log book.

the associated Blue Funnel Line, and be with the children at Easter. Flower had such detailed information that all that was needed was a decision. This was facilitated next morning, when determined to make one of the two trips, we entered the tourists' office, and were told that the Zeppelin was already fully booked. "So we leave from Rotterdam for South Africa at your birthday," was Flower's short reply. A few hours later we had the reply from Rotterdam that a cabin on the "Boschfontein" had been booked for us, and as only a fortnight was left to our departure, we had plenty of work with the preparations for the journey. My best friend was just then preparing his immigration to Chile. He tried by all means to make us change our mind and to go to Chile with him and his charming wife, who was Flower's best friend, but Flower and I preferred to go to an English-speaking country, particularly as we had some knowledge of that language.[56] The fact that my cousin was living in South Africa also influenced my decision, as I had read his enthusiastic letters at my uncle's (The first four years of my stay there in beautiful South Africa proved my decision to be right, as I came really to like this country with its freedom.[57] But during the last six months in an internment I have often thought that surely in Chile I would not have been stamped as a possible enemy of the State after having been loyal as

Above: Rosenstein's log book list of flights in his Bücker 131 aircraft – he gives an account in his memoir of how he painstakingly ushered these planes out of Nazi Germany when he escaped to South Africa just before the outbreak of World War II.

a citizen [of South Africa] for four years and after having trained more than fifty pilots for this country in my school.[58] Nobody could possibly accuse me of any subversive activities which could in the slightest justify my internment, not even as a precautionary measure under difficult circumstances.) – Again I must apologize for this digression from the subject. As the German saying goes: He, whose heart is full, will have many words in his mouth (or, in this case, in the pen).

Just a week before our departure I saw an advertisement in the "Flugsport", a journal on flying. It said that the Buecker Works[59] had had very good successes with exporting their machines[,] especially to Switzerland. As I had not thought of how to make a living abroad, I had the idea to try something in the motoring or flying line, at the same time trying to save thus part of my property by transferring it. (A direct transfer of money was, at that time, no longer permitted.) I therefore got in touch with this firm and quickly paid them a visit before my departure. I was pleased to find out at this opportunity that both the director, Mr. C.C. Buecker, and his chief designer Anderson were Swedes, who had only recently come to Germany from Sweden. I was glad to see that Mr. Buecker was very interested in my plan. With some hope, therefore, I went to Rotterdam and boarded the "Boschfontein." Apart from a few rough days in the Bay of Biscay, when we both became sea-sick, we had a beautiful trip. It was our first great sea journey and we thoroughly enjoyed it. The only bitterness was brought into it by the thought of having to return to Germany. Would no new regulations be brought into force which would make our ultimate emigration impossible? How much we envied the emigrants on our boat who, though going into an unknown future, had shaken the dust of the country, that did no longer want to have them, off their feet! We arrived up to time [sic] at Cape Town, and I shall never forget the marvellous [sic] panorama of the town and its mountains in the light of dawn. The same night we left for our destination Johannesburg, where we arrived after a train journey of about thirty-six hours. In these days it was still possible to pay the whole journey and the stay of four weeks in Germany; and we therefore luxuriously stayed at the Carlton Hotel. We spent four glorious weeks in Johannesburg which passed in no time. On the boat I had met a very charming pilot from Rhodesia to whom I had talked about my plans. He had given me information as to whom I should contact in order to find out whether my plans were practicable. I visited these gentlemen on various aerodromes and obtained many hints which became very useful later on. The result was that it was well possible to establish a flying school as there was not much competition in this field. The possibilities of selling machines seemed to be less promising. But at any rate it appeared worth a trial.

The "spirit of flying" which connects the pilots of all nations once again proved true. We were very fond of the city of Johannesburg, and a few short trips in the surroundings with friends gave us an idea of the country's vastness. The hour of departure came only too quickly. Our decision was made; we would immigrate to the Union of South Africa. We took leave from the numerous friends we had made in the short time, promising them to be back in a few months at the latest. With sore hearts we left for Germany, being thoroughly depressed by the feeling of not knowing what expected us there and whether we would be able to wind up our affairs in such a way as to be back in Johannesburg soon. The journey back on the "Anchizes" was if anything still more beautiful than the trip out, because we considered ourselves more experienced sea travelers by now. We were to arrive at Liverpool, and this would give me the chance to visit my only sister who was living at London, and to introduce her to her new sister-in-law, my dear Flower. As girls the two had already met in Stuttgart. We spent some pleasant hours, obscured only by a letter from my faithful Jörger who had gone specially to Switzerland in order to send it. He had addressed it to my wife, so that she should be able to prepare me carefully for its contents. With a few words he said that the "Zollfahndungsstelle Stuttgart"[60] (a part of the Customs Authorities whose task it is to get hold of Customs deceivers, etc.) had been informed that on my trips to Switzerland I had smuggled money out of Germany in the following way: I was alleged to have had a special pipe welded into the chassis of my Mercedes car in which the money had been hidden. The "Zollfahndungsstelle" had withdrawn their accusation "with the expression of regret." However, he felt compelled to inform me of the events with a view to the general lawlessness in Germany as far as Jews are concerned. He left it to me whether I would return to Germany under these circumstances or whether I preferred to try to wind up my affairs from England. Flower immediately said: "Under no circumstances must you go back to Germany." When I told her that she knew as well as I that all this were [sic] only lies and that I had never done anything of the kind, she tried to convince me – what with the lawless state in which Jews found themselves in Germany – a mere suspicion was enough to bring a man into a concentration camp, even though he was innocent, and that we had sufficient proof among our own friends. But I insisted on my own intention to fly straight to Berlin from London. I argued that, had I stayed in England, trying to wind up my affairs from there, this would immediately have been considered as an admission of my guilt, and would have resulted in the immediate confiscation of my property.[61] The only concession I made to Flower was that I agreed to her not accompanying me to Berlin – as was originally intended – but letting her fly to Switzerland, so that, should anything happen to me, she would be with the children and in a position to help me from there. Flower on an English plane and I in a German one left Croydon at almost the same time, and I must admit that I did not like saying goodbye to her, though, of course, I did not let her notice this. When Flower's machine left the ground five minutes before mine, I had a funny feeling in my stomach, and I wondered how long we would be separated from one another.

We had made out that, immediately on my arrival in Berlin, of which the Buecker Works had been advised, I would send her a telegram from Tempelhof Aerodrome, whether everything was all right. Flower had assured me again and again that she definitely had the feeling as if I was to be arrested out of the plane. After a landing at Amsterdam I got on to the plane for Berlin (not without having asked myself at least a hundred times: is not commonsense [sic], after all, better than courage, and should I not give up this

Above: Image of Rosenstein's flight log in South Africa – here he was flying his Piper Cub (listed at the top) and his last entry is for May 18, 1949. He was killed in a plane crash a few days later on May 23.

journey to Berlin [?]). But as I had a clear conscience I finally flew off to Berlin. My mood became worse and worse the closer we came to Berlin. But the most frightful moment came when, after the landing, the steps were rolled to the plane, and a police official came forward, doubtless calling my name. I had a vision of the concentration camp and cursed myself for my stupidity not to have followed Flower's advice. Then I replied—having no other choice – he stood at attention and informed me that…Mr. Buecker personally expected me at the dining saloon of the aerodrome restaurant and that he had been asked to take me to him immediately. It is a pity that nobody with a movie camera was present. I would like, even today, too see the expression on my face in those seconds. First the terrible fright and immediately afterwards the reaction. Never before was I so pleased with a Police official. In the politest way I was taken by the official to the dining room, without any Customs examination having taken place; and here I was shown to Mr. Buecker's table. I only wonder what the Police official thought of me, because surely he would not have treated a Jew in Nazi

Germany in such a polite way. Mr. Buecker's remarks threw some light on the matter: he had told the man to meet at the plane a certain Mr. Rosenstein who was arriving straight from South Africa for some very important discussions. But he assured me that he had only said this because he was greatly in a hurry and was afraid that the Customs and other formalities would have taken too long otherwise.

According to time-table, my wife should land at Zurich about 1½ hours after my arrival in Berlin. Having had to pass no formalities, I thus was about one hour ahead of time, of which I had to make use to save my wife from worrying. I hurried to the Post office to send the first "Blitztelegram" (a very urgent telegram) in my life, addressed [to] Mrs. Rosenstein, Duebendorf Aerodrome, Zurich, arriving from London. The contents was: "Safely landed Berlin." Now she knew everything was O.K. A telegram with the same wording I sent to her under the address of our children, in case the one directed to the aerodrome should not reach her. An enormous weight had come off my heart. I had heard and read of too many "good" things about the concentration camps, and had no desire whatever to have a change of this kind after the sea trip and the beautiful weeks in South Africa. (after all the "regret" of the Stuttgart Customs Office might have been a trick to lure me back into Germany.)

I spent the next two days with Mr. Buecker to make him acquainted with my future plans which had got more shape by now. I found much understanding with him. He had been manufacturing planes for ten years in Sweden and was highly interested in all airport business. His chief designer immediately examined my special wish to have the machines which I might buy from him fitted with English Gipsy Motors (Gipsy Major),[62] and satisfied with my visit I left Berlin and flew to Zurich, so as to spend the approaching Easter holidays with Flower and my children in Switzerland. A definite decision about the planes I only wanted to take after my return from the Swiss journey, as I always talked everything over with Flower. She had given me much good advice, something I often miss to-day. More about this later.

The flight from Berlin to Zurich was one of the worst in my life. The route was Berlin-Stuttgart-(landing)-Zurich. Half way between Berlin and Stuttgart the pilot of our passenger plane received a radio message that the Stuttgart aerodrome had "QBI," which equals a prohibition of landing because of fog and snowstorm. (Incidentally, the pilot was an old pupil of mine from my Instructor's time at the Gotha Factory, and he was very pleased to be able to take his own instructor through the air.) He informed me on a piece of paper which he sent into the cabin with the wireless operator, and told me at the same time that he had orders to fly to Munich to take fuel there, as he would not have sufficient for a non-stop flight to Zurich. This should prove to be most fortunate, because shortly before Munich one of the three motors of the Junkers plane stalled with sparkplug trouble. Thus we could just reach Munich aerodrome with the remaining two motors of the plane which carried the full number of sixteen passengers. After this experience 14 passengers were fed up and gave up the flight.

We arrived in Munich at about the time at which we would have reached Zurich. The engine failure was quickly repaired, and with only two passengers we started off for Zurich, darkness approaching. The pilot declared that this was only a stone's throw. We landed in complete darkness, and my old pupil made a perfect night landing, which I would not have done as well. I had made an appointment to meet my wife in a small place on Lake Zurich and, without incident, should have arrived there at about 5p.m. The whole family went to each train from that time onwards to meet me at the station. After the third unsuccessful attempt, the hotel manager discreetly told my wife, expressing his regret, that a German plane from Berlin had hit a mountain in the darkness somewhere near Zurich and that apparently all passengers had burnt! He said that this was reported to be a military plane but he had not been able so far to obtain event news.[63] Half an hour later I arrived and my wife, who had naturally said nothing to the children about it, was much relieved when I appeared undamaged. (Incidentally, a military plane had actually lost its course in the horrible weather. Its wireless apparently had failed, as was afterwards found out. All [the] people were dead!) It took quite a time for Flower to recover from the fright and for me to recover from the flight.

The holidays were only short. After we had told the children about Africa, they had to go back to school. When we said goodbye to them, we promised that the next time we came to them, we would fetch them for the trip to Africa. This helped to ease the farewell. I need hardly explain what it means for a child to be allowed to travel to Africa. Even to us grown-ups the first journey had been a great experience.

After that, Flower and I flew to Stuttgart. We travelled only by plane because this was the only way for us as Jews to avoid possible trouble with the German frontier authorities. I did what was most urgent for my emigration. From there we flew to Berlin to take the necessary steps which should enable me to take with [us] the three Buecker planes, which I had decided to buy in order to open a flying school in Johannesburg. The discussions with Mr. Buecker were soon finished while those with the authorities made very slow progress and nearly failed for the following reason. I had demanded that when I took these planes with us, the "Reichsfluchtsteuer," amounting to 25% of property, should be remitted ("Reichfluchtsteuer" is a tax which had to be paid by every man of means of leaving Germany.)[64] Flower and I were living at a hotel, and this is essential to make clear the events that followed. The discussions with the authorities came to no end and we began to fear that we would have to leave unsuccessfully, as we had set ourselves, as the last date for our emigration, the end of the Olympic Games.[65] For the better understanding of the following paragraph I would remind the reader that during the Verdun offensive, 1916, my observer Lt. Martin and I were shot down, and that thus Squadron 19 had one crew less.[66] My successor as pilot was Lt. Siebel who distinguished himself in the further course of the war and became a good friend of my later Squadron Leader 1st Lt. Goering. – While dining at the hotel one evening, Siebel came into the room and happened to pass our table. Since the war, I had only had one chance meeting with him at Berlin, years ago. He hesitated for a moment, for he did not recognize me immediately. Then he greeted me with old cordiality, pleased to meet an old fellow-soldier. I introduced him to my wife and he sat at our table for a short time. Just before he left, he asked me about my plans for the future, and I told him that I was making preparations to emigrate to South Africa. He fully understood this, and asked all of a sudden: "Does Hermann know about it?" At first I did not understand his question, as not a word had been spoken about Hermann Goering, and he had to put his question for a second time before it dawned on me that he was thinking of Hermann Goering. I assured him that he had not the slightest knowledge about my plans and would not hear anything about them, as he would understand that I had not the slightest connexion [sic] with him since the war. Siebel said he understood, but was of the opinion that Goering should do something for me and should at least assist me in my emigration. I replied, saying that H.G. probably had other troubles now, but Siebel left our table saying: "As soon as I see Hermann I shall tell him about you." He added: "You will hear from me."

To my wife and me that was the end of this incident. We felt perfectly sure that we would never hear from Siebel who had just talked a little too big. How great was our surprise when, scarcely a week later, I received a letter at the hotel in which Siebel informed me that he had seen H.G., who had instructed his adjutant upon his advice to write to the Ministries which had put obstructions in my way (of which I had told Siebel), telling them that I should be caused no such inconveniences. It would go too far to go into further details about the troublesome conferences at various Ministries (though it would be interesting to prove how they worked against each other). Finally I succeeded in settling everything, but I must admit that Goering's letter made things easier in some ways. I was allowed to take three Buecker planes and the necessary spare parts to Africa, and this was a privilege which was not granted to other Jews at the time (Summer 1936). The Ministry of Finance, however, did not agree to my request for a remittance of the above mentioned tax. I had to pay 40% at once, and the remaining 60% were only to be paid after two years with the provision that, should I sell within these two years a fixed number of planes for the Buecker Works, those 60% were to be remitted. At that moment I had unfortunately not recognized the importance of this clause, but was in the belief that it was something positive for us. Only much later, in Africa, it became clear to me that, if I had had to pay this tax from here, this would have meant 70% of my property which had already dwindled into a small sum by the transfer. It was only when I started in South Africa to build up my flying school and tried to sell planes to people I did not know, that I became conscious of the trap which the Ministry of Finance had [laid] out at my feet as a farewell. Fortunately the three machines I had taken with me were successful and I succeeded in selling the required quantity of planes. My firm, the "Union Aviation Company (Pty.) Ltd.," sold six machines to the flying school of the Witwatersrand Technical College alone, which gave full satisfaction. Thus the tax was cancelled.

With no fault of mine a large number of fairy tales have been circulated about my relations to Goering. I therefore wish to emphasize that everything happened exactly as I have told it in the above and that all versions different from my own are therefore outside inventions. As mentioned before I have never seen, nor spoken to, Goering from the moment of my leaving his squadron, i.e. at the end of 1917.

Shortly before the Olympic Games everything was ready for our emigration. And it was high time too! Our private and office furniture, etc. were packed in Stuttgart into three liftvans [sic] under the supervision of the Customs Office, and were sent to Hamburg. The three planes were packed at the Buecker works under my supervision, so that they were safe on the sea transport (they, too, were packed in big crates). They were taken on big lorries to Hamburg, so as to be in time for the "Tanganyika" of the German African Lines. One of the conditions for our emigration is that we should travel on a German boat. Flower and I met our children in Switzerland and drove to Italy by car, boarding the ship at Genova[67] -- It would go too far if I were to tell all about the journey. I only want to mention that we got our last bad taste of Nazi Germany on the journey which went via [the] East Coast [of Africa]. A certain Dr. Peters from Hamburg and his wife made a so-called information trip to Dar-es-Salames,[68] and they caused a most unpleasant atmosphere to be spread over the whole boat. Scarcely had they left it, [and] the crew, from the captain down to the last steward, began to breathe more freely and became quite different men. That shows how two Nazis can tyrannize a whole ship.

We went on land at Durban and drove up to Johannesburg by car. For a short period we took a flat until we had carried out our plan to buy a small house. This we did after as little as four weeks. Since then I have been living in my house, 68 Tyrwitt Avenue, Melrose, Johannesburg.[69]

From a talk several months before my emigration at my hotel in Berlin I knew that Mr. K. Katzenstein, the former partner of the Raab-Katzenstein Aeroplane Factory at Kassel, had also gone to South Africa. (I had met him before through his partner Raab who had been one of my squadron fellows.)[70] He wanted to come to an agreement for co-operation, but I could not follow his request at that time, not knowing whether my emigration would be successful, or whether inseparable obstacles might prevent it. Now, having got my machine to Johannesburg safely, I took up his suggestion to found a flying school together. His contribution to the firm which I called Union Aviation Company was the experience as flying instructor, and I brought into it my capital and the three machines. The beginning was not easy for us and cost a lot of money, especially as the undercarriage of the Buecker planes did not stand up to the strain of the South African landing grounds. After these difficulties had been overcome, an incident occurred which prevented further collaboration between the two of us, and Mr. Katzenstein left the firm. I let him go with some reluctance as I have never met a better and more conscientious flying instructor. Another instructor took his place, after a short time leaving upon his own request, because he did not have enough opportunity to fly himself. Yes, such things exist, too! More often the instructor has to fly too much. For some time now, the school was without an instructor, until at last I found a substitute in Miss Doreen Hooper. I want to say quite frankly that I have always been opposed to women flying, but if anybody had cured me from having such prejudices, then this is Miss Hooper, who, with her scarcely over twenty years, has been one of the best instructors and can well be compared with the best male instructors. With her as instructoress[71] the school developed satisfactorily, and though war had broken out by now, I looked into the future with confidence. I believed to have [sic] one advantage over other schools, namely that my instructoress would under no circumstances be called up for military service. But once again I would have miscalculated. One day Miss Hooper was informed that she had to report at Pretoria, and she returned with the sad news that she would have to leave the firm immediately because she had to join the Army. Under the prevailing circumstances no other instructor was obtainable, and I therefore decided to close down the firm and to try to sell the planes. Since the beginning of the war I myself had no longer been allowed to fly, as all licenses of alien pilots were cancelled. Here I would like to mention one more instance. I myself have not worked as instructor in South Africa. In order to obtain the South African Instructor's license, one has to pass the R-pilot's examination which included aerobatics. My firm began regular instruction at about June 1937. In the three years of its existence we have trained more than fifty pupils, of whom, according to my knowledge quite a few are to-day serving in the S.A. Air Force, and some also in the Royal Air Force. This can be considered quite an acceptable result for such a young school; even though we may have been denied financial successes, we can be satisfied with the moral success of having made our reputation as beginners. Moreover, in those three years not one of our pupils suffered the slightest injuries, which proves how carefully our training was done. We had also given proof that our aeroplanes and engines, after some small initial troubles had been overdone, did good service under the local conditions. I was sorry not to have been able to carry out my plan to have my planes fitted with English motors, as the Gipsy Major motor is too powerful and heavy and as it would have completely changed the centre of gravity.

I do not want to omit making mention of one man who during all these years in which I ran my school gave me useful technical advice, especially in the beginning when I did not yet know the peculiar conditions in this country. He is Mr. Sidney Millyard of Messrs. Air Service (Pty.) Ltd., Rand Airport, Germiston, who most readily put his long local experiences at my disposal and helped me over many a difficult situation.[72]

For the above reasons a continuation of the school was not possible. Therefore I housed my planes at the Grand Central Aerodrome, packed my spare parts, in order to store them in the two garages of my house at Melrose, Johannesburg, until I found an opportunity to sell both planes and spare parts. I put my offices and hanger at the disposal of the Airport who required them for the Military Authorities. With heavy heart I took leave of the place where, during nearly four years, I had been able to devote myself to my old love, flying. My plan was to retire to my farm "Berg en Dal" near Rustenburg, as soon as my planes had been sold, thus prematurely carrying out a project which had been in my mind for years to come. In the first year of my African stay I had bought this farm, intending to retire to it with my wife, once my two sons would be old enough to take over the firm. I may say it right now: both calculations were completely wrong. At first, my wife left me in December 1939 all of a sudden. Naturally Flower took "her" son with her. Secondly, a few days after I had closed my school, I was interned just as unexpectedly, thus leaving my child at home quite alone.

These two new buffets of fate were too much for my nerves. The loss of Flower already had caused me to fall back on the cure for everything, M…and I tried to get over the first shock, like I did at Hede's death, by taking large doses.[73] My internment completely finished me off. Only by taking largest doses could I keep myself upright, and if it had not been for the love of my son which strengthened me, I do not know what would have happened to me in Camp. I ask to be spared to tell what I had to go through the first four weeks of my internment, during which period I was only allowed to see my beloved child once, and even then only for an hour! Moreover, one disappointment followed another. I had hoped that the terrible fate of being interned would induce Flower to stand by me, but here again I was disappointed. She insisted on our divorce being carried out during my internment (I had agreed to divorce before my internment upon her urging). Thus it happened that in "The Star" of 9th July 1940 I read that I had been divorced from Flower. This new shock made the state of my nerves so much worse that the Camp Medical Officer of the Baviaanspoort Internment Camp caused my transfer to Pretoria General Hospital to be carried out on 22nd August 1940. (I had been interned on 12th June 1940.) I remained in hospital for eight weeks without being cured and was suddenly re-transferred in this condition, together with two other internees, to Beviaanspoort. I still cannot understand the reason for this sudden re-transfer, more so as the doctor who treated me had declared me unfit for Camp life. I arrived at [the] Camp in a terrible state of health, and I owe it only to the self-sacrificing care of a co-interned doctor, that I did not suffer a still more terrible nervous break-down, which would certainly have had fatal consequences. Whereas the two other comrades who had been at hospital with me and had been re-transferred have already been released for reasons of ill-health, I had to remain in [the] Camp. It took me several weeks to become able to walk again, and even to-day I am suffering from the consequences of an unfinished treatment. There must be very weighty reasons which caused such inhuman treatment!!! – Shall I ever come to know them??? –

Shortly before my return to the Camp, restrictions were imposed upon us which were caused by insubordinations in the other Camps. Again I was hard-hit by the cancellation of all visiting permits, because I could not see my son for the best part of three months. What is the good of a whole Camp like ours behaving in an exemplary manner – every new Commandant confirming this – if afterwards we are still punished for the misdeeds of our deadly enemies in other Camps? My nerves had to withstand yet a new strain. All the inmates of the Beviaanspoort Segregation Camp were transferred to Ganspan near Kimberly, which means that even after the restrictions have been lifted visits from my son will become greatly complicated and can certainly not take place as often as I would have desired. But by now I am getting used to such buffets of fate, though it is hard not to lose courage and to look hopeful into a better future.

In conclusion I wish to give a short description of my marriage with Flower and of the reasons for its tragic end. On page 10 I have mentioned how my second marriage came about, and I described the hopes I had had with regard to the education of my son Ernest. The first years of our married life seemed to prove and confirm our hopes as Ernest really took to his new mother. Our marriage, too, in the beginning built up on mutual affection (the period of Storm and Stress should be over at our age), turned into real love, at least as far as I was concerned. True, sometimes I caught myself thinking [about] whether this was not a breach of confidence toward my poor Hede, but I had the feeling that Flower, too, felt happy in our married state. Thus, for the second time I believed to have won the "Irish Sweep," and again I looked hopefully into the future, without being able to see the clouds which gathered at the horizon. As long as the children went to school in Switzerland and Flower and I were living in Stuttgart, everything was in full harmony. Small frictions between the children, caused by Peter's being more intellectually-minded as compared with Ernest's manual giftedness, were not taken seriously by me. In our youth, my sister and I had sometimes fought each other, and became the best of friends later in life. But my dear Flower seemed to take a much more serious view even of the such small incidents, as I should only learn much later. Had I only known this earlier, a word with Ernest would have sufficed to avoid such little frictions. He had expected Peter to become his playmate, and was now a little disappointed always to see him behind his books with no time for a game (Peter was a real book-worm). Then came our sudden resolve to make an end to the unbearable conditions in Germany by emigrating to South Africa as soon as possible.

Here I wish to relate an event which happened shortly before our departure and which is important from the point of view of my health. Repeatedly I had had trouble with my gall-bladder: a large gallstone which had been seen in an X-ray Photograph had caused gall colics. Now, together with Flower, I looked up my old friend Professor Volecker in Halle, with the intention to have the gall-bladder removed if he recommended it. There would just have been time before our departure. He was very pleased to hear that after such a long time I had become married again and congratulated me upon [sic] the charming wife I had found. Voelcker was not the man to use big words. Therefore his spontaneous judgement was doubly welcome. He said he was prepared to undertake the operation, which he called a simple one; but at the same time he replied upon my asking him that he could not guarantee that in the not too distant future colics in the gall ducts would not give me the same pains again. He said that only recently he had had the same experience with his wife, whose gall-bladder had been removed by a colleague as little as eighteen months ago. Naturally I was not over-keen on [sic] an operation so soon before our departure, and thus I made use of the excuse that under these circumstances there was little purpose in undergoing the operation. After all there is a certain risk in every such operation. With a heavy heart I took leave of Voelcker, who showed full understanding for my intention to leave Germany. He was one of the few University professors who had no sympathy for the "New Germany."

I wish to draw toward the end of this exposé: therefore only a few words about the events that led up to the tragic end of my second marriage. We had accustomed ourselves to our new home which Flower had chosen; we had bought our farm "Berg en Dal" near Rustenburg, where we usually spent our weekend; I had my daily occupation at the air port[sic]; and the children went to school. At this stage the first signs of friction between Flower and my son Ernest became apparent. I need hardly mention that, though reluctantly, I remonstrated with him, and requested him to obey his mother as he obeyed me. I may be spared to describe the further development of this tragedy, and I will therefore report how it concluded. One day, on my return from the air port [sic], Ernest ran toward me almost terrified, and told me that when he came home from school, the furniture of mother had just been carried away and that she had left the house with Peter without so much as speaking to him. I should soon know the meaning of this: I found a letter from Flower in which she told me in a few words that she could no longer live together with me, because, instead of taking her side, I had always stood to [sic] Ernest: in a word, she had been jealous of MY CHILD. I myself know that I am free of all the guilt in this respect, because I loved, and still love, Flower far too much to do such a thing. What I regret is that she could make no difference between the love of a man to his wife and that of a father to his child!!! But only the future show, whether there were not more weighty reasons in the form of "dear friends" or rather one "dear friend."

At any rate, this was again a hard blow, and once again I was alone with my child, though this time not separated by an Act of God, but abandoned by a beloved woman. I must say that I never had the remotest foreboding of such a development, because Flower and I had never had any quarrels. Therefore this blow came as from clear skies. Weeks afterwards she informed me where she was living and I attempted to make her change her mind. But all attempts were in vain. Thus, after much hesitation, I agreed to divorce, so as to no longer be in her way. (before this, Flower had been employed in my office for many months: I had suggested this to her as she was looking for employment but could find no position, and had hoped, though unsuccessfully, that this would draw her closer to me again. These months were the most trying times for me.) Thus, almost as unhappily as the first one, my second marriage came to an end.

I have tried in these pages to describe the events in my life the way they happened; and if the reader has gained some insight into my life and my attitude towards it, then they have fulfilled their purpose.

I, Willy Rosenstein, herewith declare the above biographical notes, containing pp. 1 to 19, to be a true and correct statement.

Internment Camp No. 2, Ganspan
P.O. Andalusia, via Border
14th December 1940

(Signed) Willy Rosenstein

(Signed Th. J. Allan Ph.D.) (Signed) Richard H. Goldschmidt

Brit. By birth
Brit. Subject by birth and Union National

(Endnotes)

1 "*Realgymnasium*" is one of the levels of high school in the German school system. *Realgymnasium* tended to focus on math, sciences and languages. The "*Abitur*" is the German university entrance exam taken by high school students. Rosenstein refers also to the "English Matric," which is the term used in South Africa for the university entrance exam, also known as the "A-levels" in England.
2 "*Neckarsulmer Fahrzeugwerke*" – Vehicle production plant in the town of Neckarsulm, in southwestern Germany (not far from his hometown, Stuttgart). There's a museum dedicated to bicycles and motorcycle production in Neckarsulm today.
3 "*Einjährig-Freiwilliger*" – one year volunteer. Rosenstein notes that "men of the better classes" (he was from the

middle-class – and his language here reveals the prejudices embedded in imperial German society) could avoid long-term conscription.
4 "E. Rumpler Luftfahrzeugbau G.M.BH." – Rumpler Air Vehicle Production Company.
5 Hirth and Vollmoeller were both pioneers of prewar German aviation. Vollmoeller was killed in a test flight crash, and Hirth survived the war but was killed in a plane crash in 1938.
6 This was the Rumpler version of the famous *"Taube"* ("Dove"), which was the mainstay for training pilots in the early stages of the war.
7 Döberitz and Johannisthal, both on the outskirts of Berlin, were, in addition to Schleissheim just outside Munich, the main locations for training pilots before and during the First World War.
8 NSU, short for Neckarsulm, produced bicycles, motorcycles and other vehicles from the 1880s–1960s. The word *"Nedel"* denotes the particular type of motorbike.
9 The Mueggelsee is a beautiful lake just Southeast of Berlin, not far from the Johannisthal airfield.
10 "B.Z. am Mittag" – *Berliner Zeitung am Mittag* – a popular tabloid newspaper.
11 Here Rosenstein crossed out his original typed word "overland" and wrote by hand, "cross country."
12 "*Gotha Waggonfabrik*" – Gotha wagon factory (misspelled here and later by Rosenstein). They specialized in rolling stock before getting into aircraft design and production.
13 Hansa-Tauben were another variation of the Taube aircraft, also manufactured by Rumpler, Etrich, and Gotha.
14 The Verdun offensive would place this in the spring-summer of 1916 – it's not clear if "Nieuport" refers to the type of French fighter that shot him down.
15 Rosenstein's phrasing and spelling here is awkward – but it seems he's suggesting that Martin was struck in the thigh. This incident occurred on April 28, 1916.
16 It's not clear which unit he is referring to here, but in late 1915, he was assigned to a squadron that flew the Fokker *Eindecker* fighter planes, *Feldfliegerabteilung* 19. He may be getting his dates mixed up here.
17 Here he's referring to *Jagdstaffel* (or *Jasta*) 27, where he was transferred on February 15, 1917. Hermann Göring, the leader of *Jagdstaffel* 27 in 1917, would have been well-known at the time Rosenstein typed this autobiography in 1940 as one of the leaders of the Nazi regime, including the head of the Gestapo and *Luftwaffe*.
18 Rosenstein's memory is a little bit off – he did not join *Jasta* 27 until February 1917.
19 He was transferred first to *Kampfeinsizerstaffel* (Single-Seater Combat Squadron) or *Kesta* 1a and then to *Kesta* 1b, which defended against French bombers launching attacks in Germany.
20 This would be *Jagdstaffel* 40 – he made his first flight with *Jasta* 40 in July 1918.
21 Usually spelled "Pomerania" in English.
22 Degelow is rather interesting. The "Stahlhelm" ("Steel Helmet") was a right-wing paramilitary organization, composed primarily of veterans from middle-class, conservative backgrounds. Many members of the Stahlhelm would enthusiastically join the Nazi Party, though Degelow, as Rosenstein notes, was anti-Nazi. He was even thrown into prison briefly under the Nazi regime. Nevertheless, Degelow maintained his rank in the reserves and served as a major in the *Luftwaffe* during World War II. See Peter Kilduff, *Black Fokker Leader: Carl Degelow – The First World War's Last Airfighter Knight* (London: Grub Street, 2009), 174-75.
23 He's likely referring to joining the Freikorps, a paramilitary militia that engaged in street fighting with communists in Berlin and, in January 1919, crushed the communist "Spartacist Uprising." Members of the Freikorps murdered leaders of the German Communist Party, including Rosa Luxemburg and Karl Liebknecht.
24 Technically, there was civil aviation – for example the *Deutsche Luftreederei*, precursor to Lufthansa –but it was indeed very small.
25 "*Regierungsrat*" – government councilman.
26 The "*Israelistische Oberkirchenbehörde*" was a Jewish religious council in Württemberg.
27 Rosenstein's use of parantheses here is confusing – but rather than correcting his style I transcribed this exactly as it appears.
28 "Salamander" shoes still exists throughout central Europe.
29 Here Rosenstein uses a dated, and racist term for someone from mixed white European and black ancestry.
30 "Unbosomed" – a bit of a dated and unusual term for disclosing one's secrets.
31 This joke is a bit obscure, but "Grosse Los" is German slang for "big time" or a "lucky win" – in English one might say "won the lottery."
32 Pythia was the high priestess of the Greek Temple of Apollo at Delphi.
33 "*Husarenstreich*" – Hussar accomplishment. "*Fliegerstreich*" – pilot's accomplishment.
34 The "pillion seat" is the cushion right behind the driver's seat on a motorcycle.
35 The Arber mountain, known as the Großer Arber, is in Bavaria.
36 Miliary Tuberculosis attacks the lungs and causes bacteria

to spread through the bloodstream to damage other organs, resulting in death.

37 Rosenstein misspells "Lancia."

38 Mercedes produced a series of evolving sports cars between 1928 to 1930, from the Mercedes-Benz S, to the Mercedes-Benz SS, and then the SSK. Thanks to Max Coolidge Crouthamel for his expertise in Mercedes history, and technical skill as he helped with scrutinizing the variable typeface quality of Rosenstein's original manuscript.

39 It's odd that Rosenstein always refers to mph rather than km/h, which would have been typical for a German pilot and driver, but he seems to have absorbed this aspect of South African culture in his short time there.

40 Rosenstein's grammar and punctuation is awkward, but kept intact here out of faithfulness to the original text.

41 Before 1930, the Nazis never gained more than 2.6% of the vote in national elections. But after the impact of the Great Depression, which intensified political polarization and a sense of crisis in an already divided and traumatized society, the Nazis made substantial gains. They jumped to 18% of the vote in 1930 and then 37% of the vote (their peak) in the July 1932 elections, drawing votes mainly from the conservative middle-classes who voted for Hitler for a variety of reasons, including various perceived social, economic and political interests, as well as antisemitism and other prejudices (notice Rosenstein's different spellings of "antisemitism" in his text). For more background by a top historian, see Richard Evans, *The Coming of the Third Reich* (New York: Penguin Books, 2005).

42 It's not clear what the date is here – Hitler came to power and created a dictatorship in the spring of 1933. Whether Rosenstein's conversations with Dr. Levi took place before or after the seizure of power is not certain, but it's entirely possible that Levi saw the writing on the wall and decided to leave before the Nazis came to power.

43 Tennicoit, or tennikoit, is a game in which players hurl a rubber ring over a net – it's played on a court similar to a tennis court.

44 Rosenstein does not give his whole name here, but cuts it off with just "St---" in the original text.

45 "*Reichstatthalter*" – an imperial governor, or steward of the region. This was a system of governorship restored by the Nazi regime for control over the different states (or kingdoms), including Württemberg.

46 The first of May was traditionally the day for demonstrations organized by the Social Democratic Party (SPD) and Communist Party of Germany (KPD) for the celebration of workers. However, when the Nazis came to power they banned and imprisoned the leaders of the socialist and communist parties, and replaced the traditional May day demonstrations with Nazi rituals.

47 "Racio-political classes" – Rosenstein is referring to the new racist curriculum introduced into the schools after the Nazis took power in 1933. Racism was the core of Nazi ideology, and racial theory was taught as part of a coordinated attempt to indoctrinate the nation's youth. By 1935, with the culmination of the Nuremberg Laws, Jewish children were expelled from public schools. For more on Nazi racial policy and education, see Lisa Pine, *Nazi Family Policy* (New York: Berg, 2000).

48 "Master" – schoolmaster or teacher

49 The "Society of Friends", or the Quakers, are a religious denomination that emphasized pacifism and ideals of equality. This small religious movement was exceptional in its leadership in activism against slavery in the 19th century. After the rise of the Nazis, they organized relief efforts for Jewish refugees.

50 This was one of the provisions of the Nuremberg Laws, which aimed to prevent "racial mixing." The law prohibited "Aryan" women under 45 from working in Jewish households, as it was widely assumed that men of property had sex with their maids. The law also reveals prevailing prejudices about women, age and sex.

51 This sentence is grammatically tortured, but Rosenstein's wife is expressing that she doesn't want to hire older servants who may be considered second-hand or superfluous by their Nazi neighbors.

52 Though readers may be surprised that the Rosensteins celebrated Christmas, this was actually quite typical for highly assimilated Jewish families who embraced this important German tradition – see, for example, the Schohl family in David Clay Large, *And the World Closed its Doors: The Story of One Family Abandoned to the Holocaust* (New York: Basic Books, 2003), especially Ch. 1.

53 "Blumen" means "flower" in German.

54 The "Boschfontein" was a Dutch passenger ship.

55 The "Anchizes" (actually spelled "Anchises") was a British passenger ship, which was sunk by a German air attack in 1941.

56 The major Western democracies including Britain, France, the United States restricted most Jewish refugees from finding asylum. Just before the outbreak of the war, Jews were limited to only a narrow range of places that briefly opened their doors: including Chile, Shanghai, the Dominican Republic and a handful of countries where many Jews trying to escape could not imagine building a future. Antisemitism was not unique in Nazi Germany,

and racist attitudes in leading democracies, including the US, meant that Jews faced hostility in many parts of the world. On the difficulties facing Jewish families, see David Clay Large, *And the World Closed its Doors*. See also David S. Wyman's classic, *The Abandonment of the Jews: America and the Holocaust, 1941-1945* (New York: Pantheon, 1984).

57 South Africa was at this time under a system of racial segregation, which after 1948 would be known as Apartheid. This divided white European colonial settlers from Africans, who were perceived and treated by the white colonial government as racially inferior. Though as a Jew he was treated as racially inferior in Nazi Germany, Rosenstein was defined as a citizen with rights in South Africa until shortly after the war broke out (before he was imprisoned as a foreign alien) The ironies and contradictions of history abound.

58 In Rustenburg, South Africa (near Pretoria and Johannesburg), Rosenstein ran a flight training school adjacent to his farm where he trained pilots who would eventually fly with the South African Air Force and the Royal Air Force against the Nazi regime in World War II. One of those pilots included his son, Ernest, who was shot down and killed in April 1945 flying a Spitfire in an attack on Nazi anti-aircraft positions in Italy.

59 Bücker was an aircraft manufacturer founded in 1932, which made a series of aircraft, including the Bü 131 and Bü 133 biplanes, which were successfully used for training fighter pilots.

60 "*Zollfahndungsstelle Stuttgart*" – customs investigation office in Stuttgart.

61 By 1935, the Nuremberg Laws had stripped Jews of basic civil rights, so they could not defend themselves against arrest and harassment. They were no longer defined as German citizens. Further, the German government, with enthusiastic support from 'Aryan' citizens, began a policy of 'Aryanization,' which entailed seizing homes, business and other property from Jews. See, for example, Götz Aly, *Hitler's Beneficiaries: Plunder, Racial War, and the Nazi Welfare State* (New York: Picador, 2008).

62 Gipsy aircraft engines were produced by De Havilland for their various types, including Gipsy Moths and Tiger Moths.

63 The grammar here is a bit tortured, but verbatim from the original source.

64 "*Reichfluchtsteuer*" – *Reich* Tax for Fleeing: this was a tax imposed by the Nazis on Jews as they fled to other countries. It enabled the Nazis to legally rob Jews of their property in the years just before the outbreak of World War II, before the Nazi regime began implementing the "final solution."

65 The Olympic Games took place in Berlin from August 1-August 16, 1936.

66 Rosenstein is referring to his time with *Flieger-Abteilung* 19, where he served from March 1915-February 1917.

67 "Genova" is more widely known as Genoa, Italy.

68 "Dar-es-Salames", or Dar-es-Salaam, is a large city on the Swahili Coast, East Africa (present-day Tanzania).

69 The actual address is spelled 68 Tyrwhitt Avenue, which he spells correctly in the front page of his flight log – see photo.

70 The Raab-Katzenstein airplane company lasted only from 1925-1930, producing a handful of aircraft designed by Gerhard Fieseler before he founded his own company. Raab was an old comrade with Rosenstein at *Jasta* 40.

71 Rosenstein's use of the word "instructoress" may seem a bit odd, but in German one would gender a noun like "instructor" with "in" ("Lehrerin," for example), so it's natural that he thinks an "ess" is needed.

72 "Messrs. Air Service" – I can't find a record of this.

73 "M" is his thinly-veiled code for morphine.

Introduction to Erwin Böhme, *Briefe eines deutschen Kampffliegers an ein junges Mädchen (Letters from a German Fighter Pilot to a Young Girl)*

Erwin Böhme's collection of letters was assembled and edited by Prof. Johannes Werner, a Protestant theologian and historian. Werner worked on this volume in the late 1920s -- it was published in 1930, when the Weimar Republic had just plunged into the disaster of the Great Depression, which would push Germany's first democracy towards collapse and the Nazi seizure of power in 1933. Werner's research on Böhme is likely what brought him into contact with Oswald Bölcke's family and led to Werner's famous biography of Bölcke.[1] Werner selected the letters to be included in this volume. His preface makes clear his aim to promote nationalistic ideals of heroic sacrifice, and he hoped that book would be a memorial to Böhme's life. Despite his efforts to edit the volume with these goals in mind, Werner also notes correctly that unlike many published pilot memoirs, these letters are extraordinary documents straight from the front. They were not produced for mass-audience consumption. Rather, they are intimate glimpses into the relationship between Erwin Böhme and his fiancée, Annamarie Brüning.

These letters offer a fascinating look into the experiences of two witty and intelligent individuals who candidly share their thoughts on the war. Their letter exchange reveals counter-balancing forces that many young people must have endured during 1914-1918, as their lives were engulfed by loss of friends, shortages, and psychological stress, but also the emotional refuge of falling in love. Most of the letters provided here were written by Böhme, but the letters from Annamarie (whose last name is only given as 'B')[2] are just as engaging, especially for students and scholars interested in life on the home front. Annamarie worked in a War Relief office outside Hamburg, serving as a nurse and helping disabled veterans and their families with the bureaucratic challenges of applying for assistance. She thus saw first-hand the traumatic impact of war as she visited destitute wives and children in Hamburg's slums. Expressing great pride in contributing to the war effort, she described herself as providing both practical and emotional support for men and their families. Interestingly, Böhme was very supportive

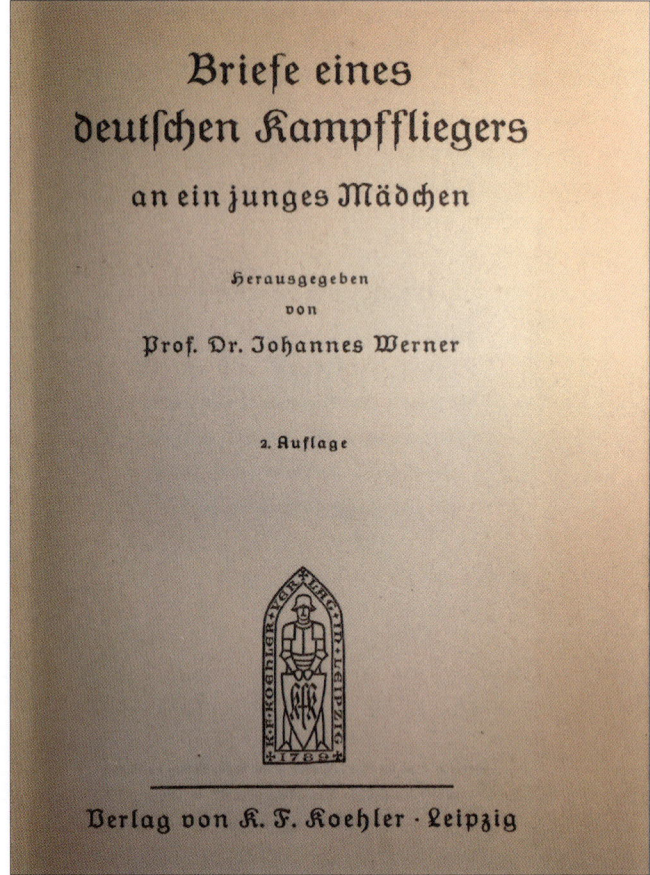

Above: Book frontspiece, Erwin Böhme, *Briefe eines deutschen Kampffliegers an ein junges Mädchen.*

of Annamarie's war work, and when she complained about the stress of her job, he expressed empathy and suggested that the stress she experienced was comparable to his own, enabling them to understand each other better. This was quite unusual, as men often complained bitterly that their wives and girlfriends 'distracted' them, or that the problems on the home front were not as important as the strain endured by men on the combat front.[3] Böhme found in Annamarie someone he related to and respected, a person with whom he could share his anxieties and emotions. At the same time, while he expressed respect for Annamarie's individual accomplishments and for being 'smart,' independent, and resourceful, he did indicate in a letter (from November 10, 1917), that he believed the changes for women brought on by the war should only be temporary. "What is required by necessity should not become the law during normal times," he wrote, reflecting the view of many men that after the war women should return to the home and domestic duties, as was the norm for middle-class women before 1914.[4]

One of the things that Böhme liked most about Annamarie is that she asked about what he was going through, including his experiences flying, his loss of friends, and his insecurities. He treated her as a partner with whom he could share his emotions, rather than someone who had to be shielded from experiences they could never understand, as was so often the case during the war.[5] The couple shared a fierce sense of humor, including sarcasm about the powers-that-be. This could often be particularly caustic, and though it may be surprising that it got through censors, censorship was actually more lax that one might assume.[6] On August 15, 1916, Böhme criticized the "naysayers in Berlin" for not recognizing the importance of aviation, and a few weeks later he lumped the "higher-ups in the military administration" with war-profiteers and other "creatures" that he couldn't stand. Though much of that was couched in sarcastic humor, he made darker jabs at the stupidity of this "murderous" war, and he expressed general resentment towards people at home who he claimed didn't pull their weight or understand the losses endured at the front. Annamarie seemed to understand that venting was the best coping mechanism as she shared her frustrations with (and nicknames for) petty bosses and charlatans who tried to take advantage of the vulnerable. She often did this with a wicked sense of humor, and she cheered up her boyfriend with stories about joining her five housemate girlfriends in the bathtub on a cold winter Sunday morning, where they wrote their boyfriends, drank wine, and tried to forget the reality of war.

The most traumatic experience that Böhme shared with Annamarie was his loss of close friends, including Oswald Bölcke, Germany's most celebrated ace before the ascendancy of von Richthofen. Böhme was Bölcke's closest comrade, and when the latter was killed in a mid-air collision in October 1916, Böhme was devastated. His letters to Annamarie clearly indicate that this was a turning point in the war for Böhme, as he was traumatized by his inability to help his friend or control fate. Böhme wrote on November 12, 1916 about how he was "outwardly" able to carry on, but that he couldn't get out of this mind the impression of his best friend's death. As with many individuals who suffered traumatic psychological injuries (what the British called "shell shock," or what German doctors called "war neurosis"

Above: Painting of Erwin Böhme, March 1917 in Düsseldorf, by Prof. Hans Kohlschein

or "war hysteria" – symptoms that are now widely connected to what's known as "Post-Traumatic Stress Disorder"), he blamed himself and was nearly paralyzed by survivor's guilt.

As with many other front-line pilots, the language of nerves permeates Böhme's writing.[7] Perhaps because these were letters that he believed would never been seen by anyone else, he was exceptionally candid in his admission that his own nerves on the verge of collapse. He wrote on May 8, 1917: "For my part, I feel pretty much like an old wreck now. The rest of my resilience is shot to hell, and for the first time I feel like I got 'nerves' during the war." He also admitted that he'd become jaded. Annamarie expressed worry that he was becoming callous and desensitized, permanently changed after being immersed in so much violence. Even before Bölcke's death (in a letter from October 18, 1916), Böhme complained about the "age of mass murder by machines, technology, and chemistry, which is what modern war has become," and he hoped that pilots, flying individual combats above all this carnage, were immune to the dehumanizing effects of the war. However, these hopes gradually broke down as the war unfolded.

While many front veterans relied on religious beliefs to help them cope with the physical and psychological destruction that surrounded them, as was the case with pilots like Heinrich Gontermann (whose letters are translated in volume one of this series) Böhme only mentions religion once.[8] Though he indicated that he was a believer, he was "not a church-goer" (see November 5, 1917 letter), and he was not particularly impressed with his Protestant pastors. Instead, like other pilots and front-line soldiers, he took greater comfort in the comradeship of men who shared his experiences, and he describes in his letters his close friendships with von Richthofen and other pilots. Further, he seemed to find spiritual communion with the machines that were such an integral part of his life. He described forming "psychological bonds" with his Albatros fighter plane (October 4, 1916), as though his mind, body and spirit were unified with his machine. Even when he flew a two-seater Albatros C.III, he imagined his aircraft as having a supernatural presence. He painted a dragon on the fuselage and called it his "guardian spirit" (see August 3, 1916 letter). The use of talismans and superstitions by young pilots to help them get through the experience of war was widespread, Böhme acknowledged. It was not uncommon for men to anthropomorphize or even ascribe spiritual power to their aircraft, on which so much depended.

Böhme's correspondence reveals a complex, multi-layered individual. While he held up the image of duty, patriotism and sacrifice, his inner thoughts, especially as the war dragged on, were much more dissonant. While on leave in the Bavarian Alps, where he tried to recuperate and come to terms with the loss of Bölcke, he confessed to Annamarie (see January 16, 1917 letter) that dangerous thoughts crossed his mind: "On my lonely hike, it was very strange, but heretical thoughts came to me. When I'm out on the field I think of nothing but war and my obvious duty to harm the enemy as much as possible. But here, removed from the war, in this divine peace of the mountains, the thought came to me: isn't this mutual, endless slaughter actually madness?" He continued with reflections on the fact that he was killing people who were much like him in a war that seemed to make no sense. Böhme's ruminations seem all the more poignant just over a hundred years later, when the carnage of war between Russia and Ukraine must be producing similar thoughts in another generation of young people who are shocked that such destruction could be witnessed in Europe again.

Once again, for extensive biographical background on Böhme, I would highly recommend Lance Bronnenkant's excellent *Blue Max Airmen* (volume 11),[9] which contains extensive data about Böhme's personnel record, biography and aircraft. But here's some brief background to help contextualize this volume of letters: At 37 years old in 1916, Böhme was quite a bit older (and self-conscious about it) than most of his fighter pilot peers, who tended to be in their early twenties. Böhme was born in Holzminden (Saxony) in 1879, and he had built up considerable experience before the war. He was an accomplished skier and mountain-climber, yet he still found time to complete his degree in engineering. In 1908, he travelled to German East Africa (present day Tanzania) and worked for several years for a civil engineering firm overlooking the construction of a railway in the Pare Mountains of Taganyika, and he worked for a timber company that sent wood to the Hubertus Mill just outside Berlin. The Hubertus Mill features prominently in his letters, because in May 1916 he flew with several friends in a Gotha G.I bomber to visit friends, including the director, Brüning. There he met the director's daughter, Annamarie. That chance meeting sparked a regular correspondence, shy romance and, by October 1917, an engagement between the two.

In addition to the friendship and love he developed with Annamarie, Böhme's letters document his experiences in *Kampfstaffel* [Battle Squadron] 10, where he flew two-seaters and then eventually fighter aircraft, including the Fokker Eindecker and eventually Halberstadt and Albatros fighters. Oswald Bölcke chose Böhme for the newly formed *Jagdstaffel* [Fighter Squadron] 2 in August 1916. They cemented a close friendship before Bölcke's death at the end of October, in a mid-air collision in which Böhme's undercarriage accidentally brushed the top of Bölcke's wing, causing enough damage to send the latter to his death. Böhme continued on with *Jasta* 2, gaining success against British fighters and observation aircraft. He was promoted squadron leader of *Jasta* 29 in early July 1917. During this period he suffered minor wounds, which are recounted here, but also endured the deaths of other close friends, including Werner Voss. Böhme was killed on November 29, 1917 in combat with a British Armstrong-Whitworth two-seater of No. 10 Squadron RFC. He ended the war with 24 confirmed aerial victories. This earned him the *Pour le Mérite*, which, according to Dr. Werner, was waiting unopened on his desk the day he was killed. He fell behind British lines and soldiers retrieved Annamarie's last letter to him from his body. That letter was returned to Annamarie in 1921. She married in 1922 and had several children.[10] Böhme had four brothers, two of whom, Martin (discussed extensively in these letters) and Rudolf, would not survive the war.

Similar to other memoirs in this series, I've provided extensive endnotes to help with context and language. I'd like to thank Doug Fant for his support in producing this

Right: Sanke card image of Gotha-Ursinius G.I bomber – the type that Böhme and his comrades flew to the Hubertus Mill in 1916.

translation. Fant translated many of Böhme's letters from Werner's volume for various editions of *Over the Front* journal in the 1990s and 2000s.[11] I reached out to Doug for his blessing, and, the generous colleague that he is, he didn't hesitate to offer advice and encouragement. This new, complete translation of *Briefe eines deutschen Kampffliegers an ein junges Mädchen* is entirely my own. I hope readers find it fascinating, and I hope that it proves useful to scholars and students.

-Jason Crouthamel

Endnotes

1. For more background on Werner, see Lance Bronnenkant, *Oswald Boelcke—The Red Baron's Hero* (Reno: Aeronaut Books, 2018), 5.
2. Her full last name, Brüning, was not discovered until years after the publication of the book. 'Brüning' is of course a famous name in German history, but she does not seem to have been related to Heinrich Brüning, who became Germany's chancellor during the Great Depression.
3. Tensions between soldiers and civilians have been well-documented. See, for example, Jason Crouthamel, *An Intimate History of the Front: Masculinity, Sexuality and German Soldiers in the First World War* (New York: Palgrave Macmillan, 2014), Ch. 3.
4. On perceptions of women's sacrifices as secondary, see Erika Kuhlman, *Reconstructing Patriarchy after the Great War: Women, Gender, and Postwar Reconciliation between Nations* (New York: Palgrave Macmillan, 2008); for an overview on women's experiences in World War I, see Susan Grayzel, *Women and the First World War* (New York: Routledge, 2002).
5. On the ways in which men often relied on wives, mothers and girlfriends for emotional support, see Michael Roper, *The Secret Battle: Emotional Survival in the Great War* (Manchester: Manchester University Press, 2009).
6. Military censorship imposed on *Feldpost* (letters from the field) might suggest that men were limited in describing experiences and emotions that may have contradicted traditional ideals. However, the Supreme Army Command (OHL, *Obersten Heeresleitung*), had to deal with on average 6.8 million letters sent every day from soldiers to the home front. The 8,000 officials assigned to censure this massive amount of mail could only monitor it superficially. Thus, soldiers' letters were surprisingly frank as men revealed their perspectives on the war. See Bernd Ulrich, *Die Augenzeugen-Deutsche Feldpostbriefe in Kriegs- und Nachkriegszeit, 1914-1933* (Essen: Klartext, 1997), 40.
7. For an overview on psychological trauma in war in the 20th century, see Ben Shephard, *A War of Nerves: Soldiers and Psychiatrists in the Twentieth Century* (Cambridge: Harvard University Press, 2001).
8. On the complex impact of the war on Germans' religious beliefs, see Jason Crouthamel, *Trauma, Religion and Spirituality in Germany in the 20th Century* (London: Bloomsbury, 2021).
9. See Lance Bronnenkant, *The Blue Max Airmen: German Airmen Awarded the Pour le Mérite, Volume 11: Bülow-Bothkamp, Wüsthoff and Böhme* (Reno: Aeronaut Books, 2018).
10. Further biographical details on Annamarie can be found in the essay by Douglas F. Vant and Dr.-Ing. Niedermeyer, "The End of an Action-Filled Life: *Leutnant* Erwin Böhme's Final Letters," in *Over the Front*, 11:1, Spring 1996.
11. Douglas Fant and Dr.-Ing. Niedermeyer provided translations and commentaries on a number of letters from *Briefe eines deutschen Kampffliegers an ein junges Mädchen*. These were released in several issues of *Over the Front*, including 5:1, Spring 1990, 6:4, Winter 1991; 11:1, Spring 1996; 12:2, Summer 1997; 17:2, Summer 2002. I would also recommend Dr.-Ing. Niedermeyer, "A Retrospective View of *Ltn.* Erwin Böhme" in *Over the Front*, 10:1, Spring 1995.

Above: The starboard side of Böhme's Albatros C.III 766/16 with the dragon marking. Although it is difficult to see, a silhouette of an airplane is in front of the dragon's mouth.

Above: The port side of Böhme's Albatros C.III 766/16 with the crocodile/dragon marking. A silhouette of an airplane is in front of the creature's mouth. The creature is different than the dragon on the starboard side (see above).

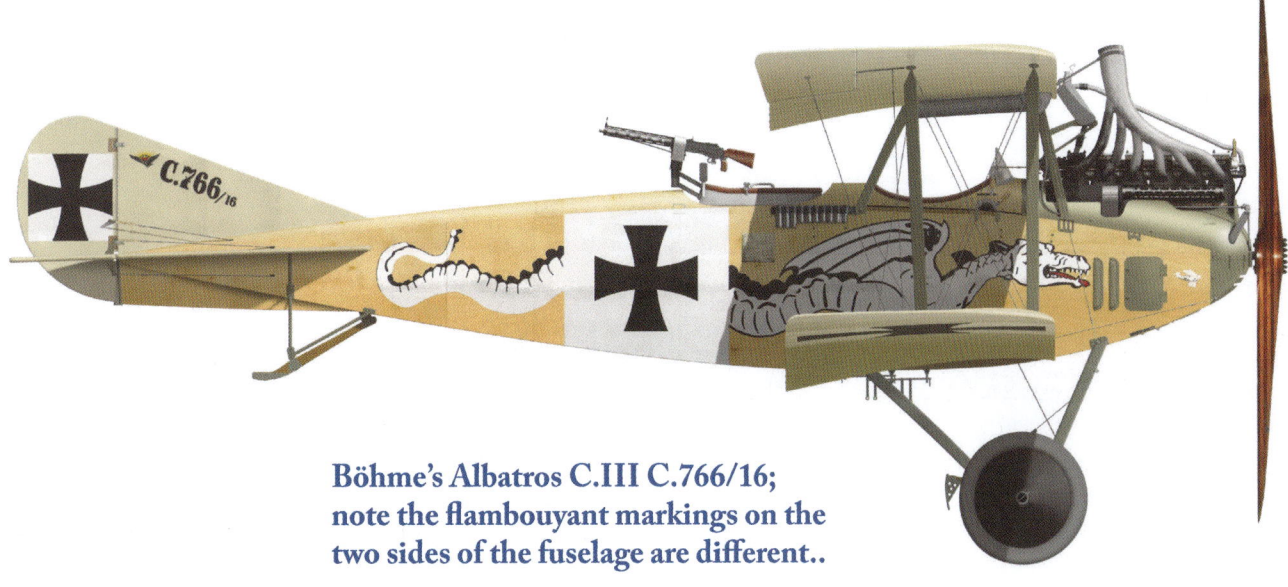

Böhme's Albatros C.III C.766/16; note the flambouyant markings on the two sides of the fuselage are different..

Engines running, a Gotha G.I is ready for take-off.

Letters from a German Fighter Pilot to a Young Girl

Briefe eines deutschen Kampffliegers an ein junges Mädchen

by Erwin Böhme

edited by Prof. Dr. Johannes Werner

Verlag von K.F. Koehler, Leipzig 1930

Erwin Böhme
Leutnant d. R. der Fliegertruppe
Leader of "*Jagdstaffel* Bölcke"
Knight of the Order of the Pour le Mérite
Fallen on November 29, 1917
Over Zonnebeke in Flanders
In his memory

Preface

A very distinctive personality, he is plain and simple as well as strong and robust. He is of a genuinely German nature: fierce while at the same time he has a deep soul and a wonderful sense of humor, closely connected to nature and at the same time extremely cultured, a war hero who tangles with enemy squadrons during the day and in the evening plays Beethoven on his beloved violin, which he takes to the front. In a word, as he would express it himself: "quite a guy."[1] This is the kind of man who wrote these letters that are so intriguing for readers.

This is *Leutnant* Erwin Böhme, who was already a well-travelled man when the war broke out, and who in his carefree youth turned to aviation. This is the man whom Bölcke,[2] when setting up the first German *Jagdstaffeln* [fighter squadrons], chose to be his first comrade-in-arms alongside Richthofen. He was the only direct eyewitness to the deadly crash of Bölcke, who had become his friend, and he was later entrusted with the leadership of the famous "*Jagdstaffel* Bölcke," and was at its head when he fell after 24 aerial victories as a knight of the Pour le Mérite at the end of November of 1917 in the Battle of Ypres.

The fact that this hero's letters are written to a young woman who becomes his sweetheart and fiancée gives them a special, warm meaning. In the middle of the war, love seized and rejuvenated the heart of this 37-year old, which had become serious and jaded over the course of his life. Our letters here contain the whole history of this love, which grew gradually, almost timidly at first, into a strong psychological attraction.

Written in 1916–1917, these letters also give a vivid picture of life as a front-line pilot and battles in the air. This picture has the advantage of being completely authentic, because the descriptions emerged directly out of his experiences and were not written for publication, as is the case of other war books.

In addition, the letters also provide his sweetheart, through their various explanations, with revealing insights into the development of combat aviation in general as well as the details of fighter aviation in particular.

Thus these letters of heroism and love form the war memoir of the fighter pilot, who flew the only weapon that was still destined to fight a knightly duel, man to man, in modern total war.

For the first section, covering the period leading up to his engagement, a number of letters from the woman he loved were also available. As a result, not only do their personalities and the development of their love come into stronger focus, but the reports of life in combat at the front also serve as an effective counterpart to the images of women's wartime well-being at home.

From the actual engagement period, there is only the one letter from the fiancée that survived, dated November 1, 1917, which Erwin Böhme carried next to his heart on his fatal flight. It was sent back from England in 1921.[3]

The desire to create a monument to Erwin Böhme, the man and the hero, from his letters has led their owner[4] to entrust them to me for publication.

Johannes Werner
Leipzig, Summer 1930

Prehistory

For six years, Erwin Böhme worked as an engineer for a German company that owned large forests and estates in Neu-Hornow on the Pare Mountains in East Africa,[5] and there, among other things, he built a suspension bridge railway for the Usambara railway leading to the high altitude town of Neu-Hornow, after which he came home for vacation in the critical month of July 1914. He was in the

Above: Ludwig Weber.

process of making big plans for travels in his beloved Swiss Alps when the world war broke out.

As an old Potsdam Guard *Jäger* [Infantryman], he immediately rushes to the flag and, as a daring man who likes to go his own way, he is drawn straight to the newly-founded air service, which was still in its infancy at the time. Despite his age – he was already over 35 when the war broke out – he was able to get himself mobilized to Döberitz, and from there he was sent to Lindenthal airfield near Leipzig for his training.

At this time he wrote to a Swiss mountaineering friend: "You may be surprised that I, having just seen an airplane for the first time in my life when I returned from Africa, went into aviation. But you, of all people, will understand this quite well: I was always more comfortable on the exposed ridge than on the rocky slope…skis and lanes seem to me to be the most dignified means of transport. Now I gratefully remember our practice at jumping on the monastery hill in Einsiedeln under the faithful care of the saint of skiing (that is, with the statue of Saint Meinrad on the hill behind the monastery). Take-off and landing are exactly the same for an airplane as for a ski-jump hill. You just have to be even more careful to only make jumps like a pro.…By the way, I've already heard a lot about the art of flying from the Marabou people in the Gonja woods in East Africa."

Although he was the oldest student pilot at the time, Erwin Böhme was destined for aviation with his youthful elasticity, his fearless boldness and his unshakable calm. As the first of a new group of people entering into aviation, he quickly completed the three exams. To his irritation, however, he was held back for almost a year as a flight instructor in Leipzig, where his pain was eased with concerts at the *Gewandhaus*,[6] and it was not until December 1915 that he managed to get to the front.

Since his youth he had trained in every noble sport, and he was an especially excellent speed skater and swimmer. At a competition on July 30, 1905, he won the "Championship over Lake Zurich" by covering a distance of three kilometers in 52 minutes and 30 seconds. During the three years of his stay in Switzerland, Böhme, as the only foreigner to belong to a select Swiss mountaineering and skiing association, also developed into an outstanding alpine climber.

From Switzerland he went to Africa. "The mountains were high enough, as Böhme put it,[7] but the country was too narrow for him. He longed for more freedom. Based on the letters of the African researcher, Dr. David, from Basel, which he read while at a friend's place, he made the decision to move to the dark continent to lead an independent life as a researcher and a hunter. He already had agreed with Dr. David that he come to him, when the news came that the researcher from Basel had died in March 1908. Since this plan had come to nothing, Böhme turned to a German company and went in their service to Neu-Hornow in German East Africa."

Nothing is perhaps more characteristic of Erwin Böhme than the route he chose for his journey from Bern to Genoa to his ship: as a solo hiker over the Rottalsattel to the summit of the *Jungfrau*[8] and down into Valais, then from Zermatt, also without a guide and all alone, up to the Matterhorn and, after a temporary camp in the rocks, he takes a descent on the Italian side to Le Breuil. So, over ice, snow and rocks, he began his journey to equatorial Africa – a man of his own strength who likes to take his own path.

The forests of Neu-Hornow also delivered cedar wood to Germany and other regions, which was prepared in the Hubertus Mill[9] for the pencil factories in Nuremberg. From this initially purely business relationship there developed a friendly relationship in 1914-1915 between Erwin Böhme and the director of the Hubertus Mill and his family.

The correspondence that resulted from this friendship offers a glimpse into Böhme's first experiences in the war:

Saarbrücken, December 22, 1915, *Kampfgeschwader* [Battle Wing] 2
Dear Director! For days I have been so far removed from time and space that I have had to let my calendar remind me with horror that it is the last date to write if my Christmas greetings are to reach you on time.

I've led a kind of gypsy-like existence since I got here. Our glorious squadron lives, always ready for action, with man and mouse in a very long railway train, which consists of nothing but captured sleeping cars – so it's just a little bit cramped but with all the comfort. In the summer it might be quite entertaining, but it has its shortcomings in the winter. Well, when you feel like want to wish for something else, you just think of the trenches and dugouts, and then you feel like you're in paradise here. Only one's shoes are in bad shape, because the train is full of young dachshunds who ruthlessly eat up or at least chew on all the boots and slippers that are lying around.

I haven't even gotten to the enemy yet. We're available here on very special orders from the Supreme Army Command and just have to curb our courage.

Our base at the moment is Saarbrücken. The landscape offers wonderful pictures from above: the Vosges, the Black Forest, and when the air is clear you see the long chain of alps – while you're up there it's easy to forget that little people are bickering below...

Mörchingen, January 14, 1916, *Kampfstaffel* [Battle Squadron] 10
Dear Director! My thanks for your greetings from Christmas time with all the friendly signatures of your people that come from the famous "little garrison." In order to reach you from our living quarters on this train, however, I need to wade through mud uninterrupted for half an hour. Here it became clear to me for the first time what is meant by the expression: the "outskirts" of the city.

Our *Kampfstaffel*, which moved here from Saarbrücken, has the task of protecting our heavy artillery, which is located nearby at the front, from the harassment of French planes. So as soon as the clouds have only a few holes, there is no lack of work.

There is a strange mixture of feelings on patrol flights, which often last for several hours, at high altitudes over the front. I climb up to fight, but when I'm up there it sometimes happens that my own situation disappears completely from my consciousness and that I only perceive everything that is visible as an image. However, that's amazing enough. From a height of 3,000 meters you can no longer see the need for a trench. But one can still see, just like on a large map, the various lines of field fortifications on either side of the current combat front and thus you get a very good picture of the individual reserve lines in this war of movement. One can also see where things went particularly badly, with increasing numbers of shell holes, which make the fields look like a lunar crater. And relatively close to both sides of the front you can see the signs of the usual operations: smoking chimneys, rolling trains, etc.

My joy is always the chain of the alps, which are easy to see when the air is reasonably clear – the peaks of the Bernese Oberland, the group of Jungfrau peaks, the Finsteraarhorn, which I'm familiar with, are particularly easy to pick out – sometimes you can see all the way up to Mont Blanc. My observer, Dr. Sander, is also an enthusiastic alpinist. He's done the Matterhorn, Dent Blanche, etc. Sometimes we both completely forget that there is a war down below, and we argue about the names of the individual mountain peaks, until we suddenly find ourselves in the midst of the amusing clouds of shrapnel from the anti-aircraft gunners and are now reminded of reality and "duty" by a rather unfriendly burst. Then we zigzag out of the shooting zone – you wish you had spurs to kick more revolutions from the engine, and you think about your sins and even more about what makes life worthwhile. Incidentally, the anti-aircraft guns reach up to an altitude of over 4,000 meters, but by skillful flying with all kinds of feints you can easily evade their precise targeting.

You will be interested to know that the famous Bölcke

also belonged to my squadron, but now he works somewhere else. His older brother is the chief of my squadron and my train compartment mate.

Landres, March 26, 1916, *Kampfstaffel* 10
Dear madam!
Thank you very much for the pretty little picture of your Hubertus Mill. That peaceful idyll is, of course, a stark contrast to the world in which I now live.

After three weeks of carrying bombs from Metz to the railway junctions of Bar-le-Duc, Ligny and also to Verdun, we are now getting closer to Verdun. We live here between Landres and Marville in our comfortable apartment block (we even have a piano!) next to an abandoned coal mine in which we have set up what looks like an extremely feudal "hero's bunker"[10] as a mess hall. The villages in the surrounding area look like Pompeii. A washerwoman who was supposed to darn my stockings can no longer be found anywhere.

We are now entrusted with a new task that appeals to me more than wild bomb throwing: "obstruction flights" [*Sperrflüge*] in front of Verdun. I guess I have to explain that word to you a bit. Our infantry and artillery have often had to gnash their teeth when French planes harassed them from above and directed enemy artillery fire at them. But we're preventing that now by constantly flying our planes up and down the front in order to hold back the enemy aircraft. Of course, there are often hostile confrontations – but most of the time, of course, they peel away from us when they see several of our planes together. Sometimes the hustle and bustle in the air is just too awesome.

Things are a little more turbulent here than in Alsace, and there is no leisure time for meditative landscape studies. The mutual surprise and methods of attack are becoming more and more sophisticated, so that the sharpest attention must not be lost for even half a minute. That's a bit stressful for the nerves during hours-long flights – luckily my nerves are strong. By the way, the most annoying thing is not the enemy's defensive fire, but the commotion produced in the wake of our gigantic projectiles as they travel through the air – sometimes they throw you around so much that you lose sight and hearing.[11]

The view of the broad battlefield from high above is strangely interesting. The sight of the burning towns and shooting batteries loses almost all of its horrible aspects, since one cannot see the effects in detail from this great height. The interrelation between the many details and the larger, complete picture of this gigantic battle is all the more incredible.

So far I have always returned safely – only my bird

Above: Sanke card image of Oswald Bölcke.

has suffered a few holes. Recently, I almost reeled in a big Farman biplane,[12] but unfortunately my observer had frozen hands (something like that could easily happen in the freezing cold at the beginning of March at 3,000 meters altitude) and the machine gun constantly jammed. While I was still chasing the Farman, one of the very fast Nieuports suddenly came at me from the lower left and put a few bullets into my wing and fuselage, but then he peeled away when I started to dive steeply towards him. In the same deceptive way, that is without being able to shoot, I then chased away another Frenchman on the way back as he was pursuing a somewhat hard-pressed stablemate of mine.

But now I've installed a second machine gun in my Albatros for my own use, so that I'm not so completely dependent on my observer and can also support him.[13] In addition to my biplane, pretty soon I'm going to get one of those nimble little Fokker single-seaters with which Bölcke and Immelmann achieve their fabulous successes. I'm really

Above: Sanke card image of Manfred von Richthofen.

looking forward to getting the new machine.[14]

Recently *Oberleutnant* Bölcke came by with his Fokker, which had a lot of holes shot through it. In the evening we brought him back to his camp near Montmédy in two cars with his motorized apparatus in tow. On the way we encountered endless troop columns, including Uhlans – but they weren't the real ones, or at least they didn't have small children on their spears.[15]

There's been storms and rain since noon today. This is called (in contrast to "flying weather") "pilots' weather," because pilots can sleep in on these days. Even the noisy heavy artillery takes a break today.

So you have a big family celebration on May 20th. I'll try to bribe my "Franz" (that's what pilots call all observers, while we pilots have the nice name "Emil" – why, no one knows), so that he can really get lost[16] and manage to lead us to the Hubertus Mill rather than over Verdun…

•••••

[Narrative by Dr. Werner] On May 20, 1916, the silver wedding was celebrated at the Hubertus Mill. In the afternoon a large military plane flew in from Berlin. It circled more and more closely over the Hubertus Mill and lured the party guests out of the house. The nearby meadow where the pilot had chosen to land was swampy – so the Ursinus biplane landed with a head stand.[17]

Emerging from the only slightly damaged plane the following appeared safe and sound: *Leutnant* Erwin Böhme, who was on vacation at the time after the difficult defensive flights over Verdun, with his younger brother Martin, also a pilot, and his comrade Ludwig Weber, who had piloted the plane.

The whole thing was a bold piece of flying – Erwin Böhme knew about the celebration day at the Hubertus Mill and wanted to personally deliver his congratulations. The surprised guests were overtaken with cheers, and the Böhme brothers even stayed until the next morning while Weber flew the quickly repaired Ursinus back to Johannisthal.[18]

This "emergency landing" had a decisive impact on Erwin Böhme's life.

Because during these days he also met the eldest daughter of the silver anniversary couple for the first time: Annamarie, whom he had not yet met during his visits to the Hubertus Mill. Inspired by the desire to deliver useful war aid, she signed up for the Women's Social School of the Inner Mission[19] in Berlin soon after the outbreak of the war, and had just completed her practical training in Vienna. Now she was staying at home for the month in which her parents' silver wedding anniversary took place, in order to take up a permanent position in the "Hamburg War Relief Organization" on June 1.

Letter Exchanges Before the Engagement

Landres, June 14, 1916
Dear esteemed gracious lady! That was such a present surprise: your friendly greeting card, which just reached me on the morning of Pentecost and immediately made my heart happy! Thank you very much for that!
Such cheering up is something that I need in this bleak, genuine "Hamburg weather," which has been plaguing us for two weeks, condemning us to do nothing: rain, hail and such a chill that we sat by our cheerful, flickering fire and sipped *Glühwein*[20] in our newly furnished summer apartment, before the expected dog days of summer arrive. As far as accommodation is concerned, we aviators can endure the war even in bad weather. But this inactivity is unbearable.

Auf einer Kuh gelandetes Flugzeug.

Above: Photo of cow struck by an Albatros during landing.

The weather finally seems to be changing this morning. I've just aired out my little bird a little bit – he was completely damp during these wet days.

The two days at the Hubertus Mill were too nice! How good it was that the little accident happened when we landed, forcing us to stay longer! – otherwise it might have just been a short, quick visit. It was such a pleasure for me to finally get to know you personally after your lovely sisters had already told me so much about you.

My brother Martin recently wrote to me enthusiastically about his stay at the Hubertus Mill. He is now in Gotha and is using his knowledge as a physics major to install a radio telegraph system in a 42-meter wide gigantic aircraft, which is intended for long flights and can accommodate I don't know how many crew and a whole store full of bombs.

Can I hope to hear from you soon? I would love to know how you like Hamburg and your new job.

Did you see the greetings from the Skagerrak warriors[21] the other day? It must have indeed been, as one has heard, a very intense and rough event. But it's also amazing to see what our blue-jacketed[22] friends read over there. It's very funny how the two English admirals congratulate each other on the "English victory."[23]

                Best regards from your very dear friend,
                E. Böhme

Hamburg, June 18, 1916
Dear esteemed *Leutnant*! I must be really proud that I received a letter from a real front-line pilot in reply to my Pentecost greetings! As a thank you for this Sunday joy that you have given to me, I thought I should also send a little letter to you while you're at the front.

I feel sorry for you because of this bad summer weather, which is so disruptive to your flying activities. Maybe it's a consolation for you that it's no better here – we're sitting in our office wearing knitted wool jackets with our fingers clenched over our files because we don't want to heat the room in June. It's good that you saw better days during your vacation! Otherwise you would not have been able to complete your hiking tour through the "Royal Saxon

Above: Bölcke's grave in Dessau.

Switzerland"[24] and you would have had even less of a chance to make your May 20 flight to the Hubertus Mill. That was such a wonderful surprise for us at the time. We couldn't have dreamed of it, and it was all the nicer that it was a real experience that I will never forget!

Think about it: now I've also had a chance to fly! *Herr* Weber[25] came to say goodbye and first took father and then me up into the air. It was wonderful! Out of respect for the treacherous meadow, we ended up in Saegebarth's rye field. My brother could hardly hold back his tears because he wasn't allowed to fly with us, but it was too late – at 8pm *Herr* Weber was still at the Hubertus Mill and at 4am his train left Berlin for the front.

Settling in and getting used to the job here is slow, but I'm making progress. You have to first get used to the Hamburg way of life and become familiar with the special conditions here before you can be useful.

It's a bit painful for me that I cannot work in practical war relief. Dr. Zahn first put me in the archive of the "Hamburg Society for Charity" so that I should first get to know all about this great Hamburg system of welfare, whose numerous threads, including all the war relief, come together in this headquarters. On the basis of detailed research, a file is kept on every petitioner. Holding consultation hours for petitioners, gathering information (especially these days for the Crown Princess' charity of children of war), drawing up files and reports – that's my job at the moment. I learn a lot from it, but I don't really like this office and filing work.

Unfortunately, I didn't get a chance to greet the sailors because their visit happened during our business hours. Our superior higher-up (known to us as "Aspirin") would have been amused if we had wanted to leave during office hours, even if it were for the sake of a patriotic cause. Our company is run in a very military-style – everything has to go like clockwork. There's also no lack of undeserved "sharp reprimands," as they call it in the military – that can sometimes dampen our joy a bit.

But overall I feel pretty good here. It helps a lot that I have a pretty room with a view of the Alster river and that another young girl who also does social work lives next door to me. Together we don't feel our loneliness so much; we prepare our meals together and also make our serious and interesting life quite fun.

I hope everything is going well with you!
With best regards, your Annamarie B.

Landres, June 24, 1916
Honorable *Fräulein* Annamarie! Is it actually a bit bold of me to address you like that?[26] But I've never heard you called anything else – and it sounds so good.

I give you high marks for replying to me with a letter. It greeted me with a good morning yesterday when I came back from a three-hour flight, raising my mood, which is usually good after work, to a considerable level. The postal vehicle met us just as we were driving to our bathing spot, five kilometers away, after breakfast. I enjoyed the letter in the shade of a poplar.

Despite the difficulties and the changes brought to you by your job, you still seem to be happy, and I imagine that everyone there is also totally grateful. It just doesn't fit with my image of you as a cheerful young lady who can run around so happily in the fields with your lute, but you now sit buried in files with a serious, scrutinizing expression and help to decide on people's fates – let's hope that slogans used aren't too hard-hearted![27]

I'm sorry that as a thank you for your good will and your loyal work you have to swallow bitter "aspirin" pills. This is something that I'm familiar with from my life in the military and I know how sick it can make you. All bureaucratic structures and conditions are detestable to me.

My ranting about the weather the other day was of some use. We've had the summer weather for a week now, and that's the end of the sleep time for pilots. We've had plenty to do these days: various bombing flights in larger squadrons, mostly before daybreak, and then again the understandably popular defensive flights over Verdun, which almost happen without seeing a dogfight. I keep getting through it unscathed, but my faithful bird is already badly shot up.

The day before yesterday, while pursuing a French biplane, I dropped down to within 400 meters of the outskirts of Belleville, allowing me to see all the details of the positions of the French long-range guns. But in these instances you can't expect too much hospitality, otherwise the enemy will get upset and start shooting – luckily mostly very badly. Unfortunately, my machine gun jammed at the crucial moment, and so my hunting expedition ended without any tangible results. Only my observer yelled out a few words of endearment to the escaping Frenchman – maybe he took them to heart.

Since you have now declared yourself to be a pilot with your first flight, I can count on you to be interested in the experiences of a combat pilot, and that is the only reason why I am reporting to you on such warlike affairs. You will have read that Immelmann crashed. The poor fellow's nerves have been shot for a while now – he really should have been forced to relax for a while; of course it's difficult to figure out one's limits. Incidentally, Immelmann did not die in a dogfight, but through a stupid accident when part of his propeller was torn off and his wings ripped away.

Our summer apartment with garden, farm and a dairy thrives in ever more beautiful bloom. We now have five proper cows and are in the enviable position of being able to live almost exclusively on our own produce. For example, this afternoon we had strawberries with whipped cream – lots of whipped cream! I know it's rude to write this to you, but one is gradually able to get to these things through "flattery."[28]

It is possible that I will soon have something to do in Kiel. One gets there through Hamburg, and so if I knew that you might meet me there, I could certainly interrupt my journey. By the way, as a member of the Hamburg War Relief Organization, wouldn't you actually be duty-bound to engage in this kind of work! Or not? If in doubt, your office might be able to give you advice on this.

This morning, on a very happy ride-about, my commander revealed to me that I was granted the iron cross first class. However, it usually takes a good two months before the award is given – that's a long time in war.

Warm greetings from your very devoted,
Erwin Böhme

Dear esteemed *Herr* Böhme!
Well, that's a real pilot's letter that you wrote me this time – even here it can't stop flying. I'm just settling down with my writing case on my balcony with the beautiful view of the Alster river, reading it again and thinking about how I should answer your last question – than there comes a sudden gust of wind, which takes the letter, making me chase after it. I race down my stairs in bold fliers' turns and just caught the little flier on the street.[29] The letter is now in front of me again, but to be on the safe side it's in my room on my desk, which is decorated today with a wonderful bouquet of sweet peas to celebrate Sunday.

After this exciting flying interlude, I want to write this greeting to you as soon as possible, because soon a Viennese

Above: Sanke card of Walter Höhndorf.

woman, whom I know from her hometown, will appear. We will first of all have some strawberries – unfortunately without whipped cream, you connoisseur! – and then we want to row on the Alster, when the thunderstorm that's now raging in the sky will have passed.

I would like to say a very special thank you for giving me a description of your flying activities. You have no idea what it means for us, who are with you on the front lines from afar with our whole hearts, to hear authentic details about what it's like there, especially when those details come from a source that you personally know and appreciate. This way you feel like you're in direct contact with the big events.

The past week has certainly brought you a lot of work again, as the newspapers reported on very heavy fighting around Verdun, and that the weather was good – you must have been right in the thick of it again. For your difficult flights I send my hopes that the French will always shoot really badly and that you don't have to fly so close to their outstretched iron arms[30] again, as on June 22nd over Belleville.

There's always a lot of activity in the air here, too. A large biplane was just flying above me, and in the evening it buzzes

Above: Sanke card of Martin Zander.

so incessantly in the air. It looks magnificent when such a majestic, somewhat smug zeppelin comes along quietly and such a bold bird flies above it and two or three others around it. You can almost imagine a dogfight when you watch the aerial conquerors side by side. But the *Heimat* [homeland] is so happy that it doesn't have to experience this!

Throughout our nation, there is tremendous mourning for Immelmann.

The little Viennese woman has just arrived. So real quick: I've been thinking back and forth a lot about how I could answer your last question. Of course I would love to see you if your route takes you through Hamburg! But the question of 'how' doesn't seem easy to me. Right, if only the Hubertus Mill were in Hamburg!

In any case, I'm pleased that you have the prospect of a little relaxation by taking a trip home – hopefully it will come true!

With good wishes and many greetings I'm thinking of you,
Annemarie B….

Kovel, July 7, 1916
Dear honorable *Fräulein* Annamarie!
Now I really wanted to go to Hamburg, but I must have gotten on the wrong train – at least I'm sitting here covered in dancing mosquitos not far from the Volhynian marshes,[31] which are full of storks. This is where the best crabs come from and where the Austrians are gloriously shortening their front. When we flew here for the first time the day before yesterday, we were still on the Styr – this morning we're already on the far side of the Stokhid.[32]

But first of all, thank you very much for your letter, which reached me here yesterday – I really enjoyed it. I must have looked funny chasing my "little flier"[33] down the street on Sunday afternoon! When I read your letter, I closed the window so that it could not escape me, because here on the Russian steppes it is sometimes quite windy…

Monday, July 10
It's also been quite windy for us these days – hence the long break in this letter, during which I was recently called to fly. Today it will probably be decided whether we can remain in our present position or whether we have to follow the retreating Austrians for a while. Fresh German troops have been arriving since yesterday – hopefully a different picture will emerge and the retreat will come to an end.

Linsingen was with us yesterday and said that our squadron had made a significant contribution to delaying the Russian advance at least a little during these critical days. But we also carried the hundred kilos of bombs to their thick columns and managed to ensure that their infantry only snuck through the woods during the day. When the bombs were dropped, we boldly dropped down to 400 meters altitude on the columns of troops and worked on them with our machine guns. There's always a tangled mess, like a disturbed ant hill, which is very funny to look at from above. Recently at the Styr river, where it crossed at the town of Kolky, all the horses were simply burned up. Russian planes, which could potentially hinder our activities, are no longer seen in the air, nor is a captive balloon.

You might be able to imagine after all the tense months of activity at Verdun that this means a pleasant change for us. It certainly is, as long as our engines remain loyal, purely a stroll in the park.[34] You can indulge in pilot jokes here that would be unthinkable on the Western front. A few days ago, I discovered a Russian airfield far inland. I dropped to about 500 meters and started hacking at it with the machine guns. One of them – just one – took off quickly, but the other machines waited in vain for their crews. Something like that would have been unthinkable in France.

How long we have to stay on the Eastern front is quite uncertain. At first we were told: it's only a very short time! But people seem to grow fond of us here.

I am now struggling somewhat with fate, which forced me to postpone the intended trip to Hamburg – if only because you would not have had much time to think about a solution to the extremely difficult problem of meeting me. "War Relief" imposes certain duties. If I had simply turned up, helpless and lost, what else would you have had to do except guide me through the chaos of the large seaside city. Always prepare for the worst and, until then, keep steady on the rudder. I'll take over the steering later in an emergency – it's not much different than flying.

Our trip from West to East lasted four days, but it was bearable in our comfortable train. In the evening we always had strawberry punch. If you let your eyes wander over the really wonderful cornfields, thoughts easily drift from the war to more peaceful times.

          With warm greetings from yours,
          Erwin Böhme

In Kovel again, August 3, 1916
Dear honorable *Fräulein* Annamarie!
You must bear with me that I haven't replied to your dear letter until today, but we have been living a purely nomadic life lately. These two greeting cards will at least show that I am thinking of you. You gave me great pleasure with the pretty photos of our "emergency landing" at the Hubertus Mill, which I will never forget, especially since I also discovered that you were in one of the pictures. Of course, if you want me talk a little bit about myself, then of course it's not going to be about "war." One really gets a bit rough over time and we find satisfaction in all kinds of unscrupulous adventures. But that has to be the case for things to move forward as a whole. Wherever this war-like spirit erodes and one is only half-hearted about everything, as is the case here with our ally,[35] then there is not much good that we can hope for as we face the challenges of the moment. If only we had the brave Tyrolians with us![36] But they [troops from the Austro-Hungarian Empire] are mostly foreign-language regiments, and that's why I regret that we are so dependent on the existence of the Austrian conglomerate of nations in world politics.[37] But I'd rather stop writing now, because it's easy for me to get really angry as soon as I come to this topic.

You ask how many combat squadrons are here now. We are the first and so far only squadron in action here on the Eastern front. So far, only a few field aviation units [*Feldfliegerabteilungen*], each with six aircraft for

Above: Von Richthofen and his beloved dog, Moritz.

reconnaissance and artillery observation, have occasionally used bombs. If you have read about "our combat squadrons" in the army report, it probably means parts of our combat wing, which inludes six squadron that have between forty and fifty aircraft.

We have developed our own customs for the Eastern front, and, according to many prisoner statements, we are unwelcome migratory birds over here. My bird is called "the scourge of Volynia." They've painted a ferocious dragon on him as a guardian spirit – our modern youth seem to believe in talismans again and cover their planes with pictures of elephants, storks, and the like. Well, our kite certainly makes a shocking impression on the Russian farmer.

My new observer, by the name of Lademacher, with whom I've been flying since July 10th, is a skillful and top-notch guy. In air combat, he always used to loudly shout insults – it's a good thing that the propeller made an even louder spectacle, otherwise the Russians would have aimed a whole barrage of insults back at us!

Yesterday morning I shot down a Nieuport biplane, which was interfering with our bombing operations over Rodzyze, Brusilov's[38] current headquarters, and I shot down a big Russian plane on both July 11th and July 13th. These last two made it pretty easy for me with their clumsy attack. Yesterday's fight was more dangerous – that was probably a Frenchman. Now it will likely be a few days before an enemy

Left: Dachshunds with their co-pilot.

plane shows up again.

From July 15th to the 23rd our squadron was with the army of Leopold von Bayern. The old gentleman was very interested in flying and wanted to fly to the front himself. Unfortunately, we were only able to fly twice the whole time because of the constant rain.

Yesterday Hindenburg was in Kovel. Thank God, he's now been given supreme command over the entire Eastern front. There must be something big underway again.

Bölcke is now on a "business trip" to the Balkans, which will probably also interest you. Immelmann's death left such a strong impression at the top that, in order to at least preserve Bölcke, he was temporarily forbidden to fly.[39] That's why he had to keep his nineteenth victory, which he shot down despite it all because it flew right over his camp, a secret. It is said that Bölcke has been chosen to organize fighter aviation, something he and Immelmann had been training men to do. For the time being, however, they wanted to give him a quiet job somewhere where he should "take it easy on his nerves." He protested against this: he didn't need a "cold water asylum" and didn't have any desire to be kept in a greenhouse.[40] So it was decided that he be given leave with the task of finding out about the status of aviation in Turkey and on the Balkan front. When Bölcke was ordered to breakfast on the day of his leave, the Kaiser said to him: "Look, we've put you on a leash." – I know all this from

Right: Sanke card of Lothar von Richthofen, with autograph.

Bölcke's older brother, who is still my commander.

Today a furious weatherly wind is roaring over Volhynia. It's already taken some airplane tents with it. Our tame squadron stork Adolar and his Jackdaw girlfriend cower in the slipstream of my car – they're very shy. As usual, the low barometer level brought me a little fever (souvenir from Africa!) – otherwise everything is fine.

Hearty greetings from yours,
Erwin Böhme

Left: Sanke card of Richthofen and Voss.

Kovel, August 15, 1916
Honorable *Fräulein* Annamarie!
It's with pleasure that I'm writing to your new address for the first time today: Sonnenau. I like that the people of Hamburg give their streets such pretty names. May you experience lots of sun in your new home![41] Life will be happy enough when so much young blood lives next to and with you – all brave girls who voluntarily serve their people [*Volk*] and fatherland in such difficult times. I didn't even know that there are so many branches of women's social work: war relief, career advice and job placement, care for small children, assistance for waitresses, educational work in the "*Rauhen Haus*"[42] – I send all my respect to these German girls!

Today I have some big news that I would like to report right away:

The famous Bölcke stayed here for two days on his return journey from the Balkans to visit his brother. He talked with great interest about his experiences in Turkey, and then, what interested me even more, is that he is now really setting up a single-seater fighter squadron on the Somme, for which he chooses the best, most select people who appear suitable to him. I fell asleep that night thinking:

It's a shame that you're such and old boy and not fifteen years younger![43] Such fighter-plane combat in a dashing single-seater – that would have been your thing.

Imagine my surprise when Bölcke suddenly walked up to me the next morning and simply asked: "Would you like to go to the Somme with me?" I have never called out a happier 'yes' in my life. Of all the vices, I'm least inclined to vanity, but I'm really proud of this show of confidence. In addition to me, Bölcke recruited a young Uhlan *Leutnant*, von Richthofen, a splendid man who has proved himself to be a bold and reliable pilot here and at Verdun.

But beyond the topic of my own luck, I'm happy about the fact that single-seater squadrons are finally being set up. For those who have followed this development in aviation with an open eye on the front, this necessity has been clear for a long time – except to our dear bureaucratic authorities in Prussian Berlin![44] Now begins a new epoch in the utilization of the airplane as a weapon of war. Only now will there be a real fight in the air, man against man.

So you can understand my enthusiasm, I have to explain something to you.

You know that at the beginning of the war our planes only had the task of long-distance reconnaissance. They did it so splendidly that the cavalry, whose domain it had been before, was soon considered antiquated. With the shift to trench warfare, this activity came to a complete stop. In its place came the photographic reconnaissance of the enemy's position, and the proper directing of our artillery fire by observing its effects on enemy infantry. None of this involved fighting in the air, for the simple reason that our pilots did not have any weapons suitable for this – only pistols and carbines. While the French aircraft were mostly equipped with machine guns as early as 1914, we only started to do this in 1915.

Then came the new task of dropping bombs. This made the aircraft an offensive weapon for the first time. But there was no question of an actual fight either – there was the lack of evenly matched opponents. I've been through that for long enough, and my hardened warrior heart was ecstatic every time our bombs landed right on their target – but I haven't gotten rid of the feeling that it's not particularly fair to throw bombs at defenseless opponents from a safe height. The only thing they can do from the ground is to use flak [*Fliegerabwehrkanone*]. Now there are optimists who claim that even when planes fly a few thousand meters high, the flak can still sometimes hit something. I never really believed that, but I always preached that a pilot can only be fought by another pilot.

That's what happened when the machine guns were introduced, and anyone who was there at Verdun can sing

Above: Sanke card of Carl Bolle, who succeeded to command *Jasta* 2.

about it. But we already had the new equipment that allowed the pilot to fire straight in the direction that the aircraft is moving. For a long time that was not the case and the observer, because the pilot was sitting in front of him and the propeller was spinning, could only shoot sideways and to the rear. If one considers a frontal attack to part of a fight, then it still wasn't a real fight.

You couldn't go straight at the enemy. You always had to turn around and catch him from the side. We fought bravely this way too, but one only took up the fight when one happened to encounter an opponent – the possibility of taking the fight to him was still too awkward.

This was all suddenly changed by Fokker's great invention, which made it possible to shoot forward through the propeller arc. Because of this my observer cannot only shoot sideways, but I can also shoot forwards with my built-in gun. It's a completely different fight.

Above: Sanke card of Kurt Wolff.

Now that one man can fly the plane and fight at the same time, it was time for Fokker (followed by the Halberstadt and Albatros airplane manufacturers) to start building fighter planes for only one occupant. Since these single-seaters have no observer and no bombs to carry, they are much lighter, much faster to climb and more maneuverable than the heavy two-seaters – they are the ideal aircraft intended for active air combat. Bölcke and Immelmann have achieved great success in these machines. This was no small shock to the many naysayers, especially in Berlin, who declared these light machines to be unfit for war and a mere gimmick.

There were also others including myself who were allowed to fly the Fokker single-seater in France from time to time – I felt as if I were a bird myself. The machines obey the will of their pilot so flexibly that you can risk anything with them, even fighting against the strongest aircraft. With these planes you can really hunt down the enemy – the term "*Jagdstaffel*"[45] is aptly chosen.

So far, however, only a few individuals have taken their game-stalking machines out on their own. If in the future we can use these miracles of technology, flown by the best pilots, in squadrons, or perhaps in larger combat wings, against our aerial enemies, we will not only keep them in check despite their numerical superiority, but we may also grab air supremacy.

Now you probably have a sense of my joy about the formation of the first fighter squadron and about the fact that I, as one of the first, can participate in this new phase of aviation.

But enough for today! That turned out to be a whole lecture that I gave you. But I think that you are interested in it.

So I'm going to regroup on the Western front soon. Maybe I'll lose a few vacation days. You know that Hamburg is still on my agenda. Thus I'll close this long letter by saying: Hope to see you again soon!

Yours, Erwin Böhme

Holzminden, August 23, 1916
First let me give a German greetings, my dearest *Fräulein* Annamarie! Having arrived at my mother's today, I would like to ask you straight away whether I can meet you in Sonnenau next Sunday (August 27). I hope I won't be ordered to the front sooner.

Happy goodbyes, Erwin Böhme

P.S. The good news of the successful return of the U-"Deutschland" has just raced through our little town.[46]

Hamburg, August 24, 1916
Against all discipline and "Aspirin,"[47] I write these lines in my office very furtively so that you, dear *Herr* Böhme, can receive my answer as quickly as possible. On Sunday you will not only find me here, but also my sister Elli, who has been with me for eight days. We both look forward to your visit.

In a rush, yours, Annamarie B…

Oldenburg, September 1, 1916
Dear *Fräulein* Annamarie!
With fond memories of the beautiful days in Hamburg, I send you my heartfelt greetings. I also feel happy here with my dear brother Gerhard. This morning I first helped him tend to the horses of the 13th Hussars, and then we had a wonderful ride through the pasture.

There doesn't seem to be any more war on the Western front – or at least they don't still need me there anymore.

Loyally yours,
Erwin Böhme

Bertincourt, September 11, 1916
Dear *Fräulein* Annamarie! The back and forth of the last week didn't give me a chance to write sooner, and today there probably won't be much either, as the weather is set for peak activity. If the roar of war gets too uncomfortable here, I think of the cozy days that I was able to spend recently in your and your sister's kind and charming company. Thank you for the goodness that you have shown to a warrior trapped between two fronts!

And now think about this: Although I stayed in Oldenburg for three days and then three more in Düsseldorf with the painter Kohlschein (his sister is married to my brother Gerhard), I actually still came here too early.[48] When I read about Bölcke's twentieth victory in the newspaper in Düsseldorf, I drove straight here without waiting for my orders via telegraph – but our new machines won't arrive until the end of this week. For the time being I had a discarded Halberstadt fighter patched up. Bölcke had his old Fokker, and two others are also still there, so at least the four of us are at the ready. Yesterday I flew with Bölcke for the first time. This morning we've had several encounters with huge swarms of Englishman who are being incredibly badly behaved right now – once we're completely set up, we'll give them a warning.

Bölcke's already downed his twenty-second. He's a really great guy who should be admired not only as a pilot, but also as a person – there's not the slightest trace of thirst for fame or bragging. He's always just focused on the task at hand, and with amazing nerves.

Yesterday, when I noticed Bölcke's Lifesaving Medal,[49] and a comrade told me the following story about him: in the summer of 1915, when he wasn't yet a famous man, he saw French boy, about fifteen-years old, fishing from a high quay wall. The boy fell into the deep canal and went underwater. Bölcke, from where he was, bent over and dove head first into the water, diving down for the boy. On the second dive he gets hold of him. Finally, he pulls the boy into a boat and smacks him on the right and left as punishment because the boy has been too lazy to learn how to swim. That's the real Bölcke!

The day before yesterday, in the fog, I took him by car on an extremely sensitive journey to the front, where he wanted to visit the "landing sites" of his last two opponents. As we were driving through the village of Miraumont and just at that moment of the most pleasant and quite peaceful entertainment, an English grenade whizzed through a house wall and exploded right in front of our car – miraculously nothing was broken but the windscreen. I found that to be a really irritating thing about the English – they won't allow even the most harmless pleasure without getting jealous – "Aspirin"![50]

We live here, spread out, quite nicely in extremely busy town of Bertincourt. I have to admit that I wasn't able to sleep properly for the first few nights before the concert of grenades and the constant rattling of trucks, but now it's all right – one gets used to everything. Our airfield is located in a beautiful little forest with tall old trees; there we are building our mess along with accommodations for daily readiness.

Yesterday my former observer, Lademacher, wrote to me that my old *Kampfgeschwader* 2 is likely to travel further into Romania. By the time the war is over, we'll all be gypsies.

Hearty greetings from your faithful,
Erwin Böhme

Bertincourt, September 21, 1916
Dear *Fräulein* Annamarie![51]
Today the main thing that I have to report to you is that the new fighter squadron has now been properly constituted. Our machines have finally arrived, only six at first; on Saturday we fetched them from Cambrai. On Sunday morning we tried them out for the first time, ran into an English squadron of eight large biplanes and finished them off – we each got one; only two escaped.

Now Bölcke has another one: his twenty-eighth. The calmness and grit with which he flies and fights are wonderful. Our new machines also border on the miraculous.[52] They are far better than the single-seaters that we flew at Verdun. Their climbing ability and maneuverability is amazing; it is as if they are living, compassionate beings who understand what the pilot desires. You can risk and achieve anything with them.

On Sunday evening our still incomplete hunting club had a small party to celebrate the inauguration and the first day's success of the squadron. At this moment, Bölcke took the opportunity to give me the iron cross first class, which had just arrived.

Sunday was generally a day of happy surprises. In the morning the parcel arrived with the lovely chocolate and the nice *Heidegrüss*[53] that you had sent to me in Russia – it faithfully followed me, even if it took a little bit of time to do so. And then, contrary to all expectations, my large suitcase arrived, the loss of which I had already mournfully come to terms with. In addition, nothing is missing except lots of good soap and one from each of two pairs of different-colored rubber boots. Now what do I do with this mismatched couple. Please advise me with your clever housewife sensibilities!

So you've found Bertincourt in your Andree![54] Meanwhile, we've had some experiences that would suggest it is not exactly a highly recommended place to live, and the day before yesterday we moved to our airfield one kilometer north of the village, because the English are so ruthless that they often shoot into the village with some long-range artillery – not exactly a lot, but enough that one can no longer call the place idyllic. The English have little sense for the idyllic, which is why you always find them in Zermatt and other grandiose locations. It's just unfortunate that the French put their village right in the middle of the theater of operations. They were quite nice to us, by the way. "[in French] The war is terrible for us, for you, and for all the world," was the poetic formula of consolation that we always exchanged with them.

Here at the airfield we're now housed in wooden barracks, which we've already set up in style: tables made of crate boards and plush armchairs in front of them, a stove made from an old cement barrel, next to it a mirrored cabinet, Perfer carpets and whatever else good nature has to offer. In addition, on rain-free days, there are uninterrupted feature films in the sea of air: you can watch most dogfights from the airfield, which, especially when Bölcke is on stage, is a very exciting and rewarding play .

I've just come from an almost unsuccessful dogfight – not with planes, but with flies. There are at least two million of these trusting creatures in my room – I give up fighting with them and instead make philosophical reflections on the reason for the existence of various creatures, including the following: people who profit off of food shortages, flies, Englishmen, gramophones, higher-ups in the military administration and the like.

I'm pleased that you have a free week in the Hubertus Mill. From the many, many greetings that I send you, take a few thousand there with you.
Yours truly,
Erwin Böhme

Somme, October 4, 1916
Dear *Fräulein* Annamarie!
First, to your smart question: how does a fighter pilot, who has only two hands, manage to steer his plane safely and at the same time point and fire his machine guns? It pleases me that you ask because it shows that you think about things and have a real interest in aviation. I will try to give my answer as clearly as possible.

We do not "point" our machine gun at all, since it is fixed to the aircraft and only fires straight ahead in the direction of flight. We don't "aim" the gun, but rather we aim the whole airplane. Thanks to the fabulous maneuverability of our machines, it's possible to adjust quickly and precisely towards the opponent – preferably from above; attacking from above always gives one advantage in aerial combat. As soon as I fly my Albatros directly at the enemy and line him up directly in front of me, I only have to press a button – then my gun hammers at the enemy until I switch it off again. This is the whole secret of my art, flying and shooting at the same time. The correct shooting as well as hitting thus actually depends on skillful flying.

However, since it is not a question of mere target shooting, and the opponent does the same maneuvers (at least if it is a daring Englishman like those we're facing now – the French, with whom I had to deal with at Verdun, were often too afraid to get into a fight), it is not only necessary to fly towards him, but it's also necessary to avoid the field of fire of his plane, and then quickly dive down on him again.

This is how each aerial battle is formed into wild circling over and under each other, sometimes lasting for several minutes. That might look terrific enough, but in this mass, machine warfare it is really still a genuine, knightly duel with the same weapons in which proficiency is the decisive factor: first and foremost confidence and dexterity is a prerequisite in aviation, then carnage and guts in the attack, and, last but not least, you need a steady hunter's eye and lightning-fast decision making skills to grasp and exploit an opportunity. Frivolous as it may sound, I take a sort of pleasure in fencing with a decent Englishman.

Such air battles would have been unthinkable with the heavy two-seaters. Our pilots have a completely different relationship with our lighter single-seater: it is, I would say, a personal bond with the machine.[55] You no longer have the feeling that you are sitting in an airplane and steering it, but it is as if it were a psychological bond.[56] It's like a good rider who doesn't just let himself be carried by a horse and tries to communicate his goals with external help, but he has become so completely merged with his horse that it immediately feels what its rider wants. Both have full understanding and trust in each other, and it is one will that motivates them.

One can only compare this delight of being completely one with the Albatros with the joy felt by a true rider. When I was allowed to fly a single-seater for the first time at Verdun, I felt blessed[57] in it, and I flew much more confidently and dexterously than in a two-seater.

For me, this may have something to do with the fact that throughout my life, my guiding principle has been "nobody else can stand up for you. One stands up entirely on your own." The new dominance of the air is similar to conquering the alpine peaks in that we are also led into the realms of the air that are inaccessible to most people. I didn't think about it in any other way. I have never allowed myself to be dependent on ropes, even on the most difficult climbing expeditions. It made me feel relieved that I didn't have to feel responsible for others, and it was a much greater feeling of victory when I had to overcome all difficulties and dangers with my own strength. So as an old alpine "solo climber," I was predestined to be a fighter pilot, and now I'm glad and happy to be able to help the fatherland in this way.

It is a pleasure to be allowed to perform in our *Jagdstaffel* with nothing but men who are inspired by the same spirit. Our incomparable master Bölcke chose his men with a sure eye. In the three weeks since the squadron has existed, we've already put 27 Englishmen on our list. Bölcke alone has ten of them. They already know our little Albatros fighters quite well over there and are no longer as brash as they pretended to be three weeks ago when they were cocky because of their superior numbers.

On September 23 I forced down two planes from a squadron. Of course, while I stalking the squadron, one of them was also brought down by an anti-aircraft unit stationed nearby. I don't begrudge them that – those poor guys also like to book a success for themselves from time to time.

It's terribly funny when the day's events are described in the mess each evening. You can watch a victor describe his fight: they wave their arms around in the air and talk "with their hands," so you think you're at the stock exchange.

With the longer evenings now, life in our cozy mess hall is usually quite stimulating. Guests often arrive – all our old friends meet up again one by one during the Somme battle. It's very interesting when "old guns"[58] and others, including Helmut Hirth, talk about the early days of aviation.[59] What incredible developments there have been in such a short time span! I think to myself that later, in peacetime, all this will support the development of the airplane as a means of transportation.

Unfortunately, our new circle has already thinned out. The squadron has already lost four brave comrades in air combat, including Hans Reimann, our excellent piano virtuoso. The last man who was killed, just three days ago, is my friend Philipps, whose parents were my closest neighbors in Africa – and his death was caused by a stupid accidental hit from an anti-aircraft gun; we were all in his immediate vicinity.

Your announcement that two zeppelins, the L 6 and the "Sachsen," burned up in their airship hangars, was news to us.[60] Such bad news is indeed not allowed to be published right now – similarly you only hear from the army reports and newspapers what you are supposed to hear and are allowed to hear. But that's how it has to be right now.

Will my letter still reach you at the Hubertus Mill? Your mother made me happy the other day, even before your letter arrived, with two pretty pictures taken under densely blooming lilacs, accompanied by the wish that by the time the next lilacs bloom, peace may once again reign in Europe. For us at the front, this is a question that we hardly ever vent. We just do our duty without worrying too much about the future. And by the way, the Somme would now probably be the last suitable region for entertaining thoughts of peace. First we have to win.

I greet you from the bottom of my heart and along with you the dear Hubertus Mill.
    Yours truly,
    Erwin Böhme

The Somme, October 18, 1916
Dear *Fräulein* Annamarie!
I have to give you my joyful congratulations on the new field of work that opened up when you returned. I've always had the feeling that you don't like office and filing work – it would be tormenting to me. So praise be to the excellent Dr. Zahn, who has moved you to the effective and practical job of war relief! The district of Barmbeck is to be envied now that it's under the care of your gentle and firm hand! May you work there joyfully and with blessings! How was it saying goodbye to "Aspirin"? Short and painless, I would imagine!

You admire our Bölcke. Who wouldn't?! But all of you just admire him as a successful war hero – you can't know anything about his rare personality. Only the few who are lucky enough to live with him know who he is. This unassuming young man, far from letting his head be swollen by fame, has a level of maturity and detachment of character that is downright incredible when one considers that in his short life before the war he really hadn't experienced anything special. But I assure you: I do not only admire Bölcke as my master – as strange as that may sound, since I'm 37 and he's only 25 years old – I respect him as a person and I am proud of the friendly relationship that has grown between us. I like to think that he sees me as an older mature man and he's happy that I am still totally committed to the same things for which his heart glows.

It's a very unique thing how Bölcke transmits his spirit to each and every one of his students, and how he pulls them all along with him. They go with him wherever he leads. No one would ever let him down. He's a born leader!

No wonder that his squadron thrives! We have now expanded with the addition of a few more young people. The squadron victories are piling up. Despite the many fights and some bold stunts, we've not suffered a single loss in the last two weeks.

But stop! We have yet one more loss to mourn: Euphemia, the bravest of our nourishing squadron cows. I hope you won't find it blasphemous that I so abruptly switch from serious to joking. In her enthusiasm for the noble art of aviation, Euphemia recently wanted to happily greet an airplane returning home in the evening's semi-darkness – they ran alongside each other, and afterwards they were both, cow and Albatros, all bent up. The Albatros was not difficult to repair, but Euphemia soon breathed her last bit of life that was so precious to us.

Now I actually have to talk with you about something more serious. Your letter sounds like it contains a faint reproach that we simply number our victims, and that we regard our fallen enemies as mere numbers, so to speak. To be honest, I've never thought about it and it would hardly have occurred to me that there were something that you could call, let's say, callousness behind it. Now that I think about it, it's easy to explain the pilot's need to do this once you consider the nature of our struggles.

I can't imagine that the commander of a heavy artillery battery, firing from a hidden, distant position with a large caliber gun at an invisible enemy, could ever think of calculating: "Well, this bomb has once again shattered or buried such and such a number of men." If you compare our fighter pilots with this kind of long-distance combat, in which the opponents never see each other at all, then the difference between the two and thus the peculiarity of our fighting becomes clear to you.

With us, every fight is a personal battle, man against man, with the same weapons and equal opportunities. That's the wonderful thing about fighter aviation, that in this age of mass murder by machines, technology, and chemistry, which is what modern war has become, we are still the only ones fighting an honest man's fight, face to face with the enemy. Consider that when I circle around my opponent in a whirlwind flight, I recognize him precisely. Sometimes I get so close that I can look him in the eye.

Every fight here is a joust, a knightly, or to put it in more modern terms, a sporting duel. I have nothing against the individual man that I fight against. I just want to put him and his plane out of action so that they can no longer harm us. If my opponent has to make an emergency landing in our area, I'm happy to shake hands with him afterwards. If he's fought honestly and bravely, I respect him – but he's still booked as a "number." As you can see, there is no contempt for our opponents when we do this.

The fact that one books his victories is probably now more understandable: it's because every fight is a personal one. You also credit your success personally, and if the successes increase, you just have to number them.

In the past I often smiled at the fact that the Couleur students counted their duels so conscientiously: how often they were "cut" or hit. I can understand that now.[61] Their duels and ours have a lot in common – except that we always deal with a matter of life and death. We are also alike in that in the end we do not fight for ourselves, but for our colors. Every success I get counts not only for me but also my squadron, and we strive for success in service to it. It's called the spirit of the corps.[62] This dominates especially in *Jagdstaffel* Bölcke, and our splendid young people are fired up for it. It might seem odd that an old guy like me would get so carried away by it – but it's a sacred fire.

After this lengthy speech, the undersigned has the honor of hoping that we will soon continue to number our victories further into the future – hopefully quite a lot very soon!
Write to me soon about your new job!
Yours with loyal greetings,
Erwin Böhme

Lagnicourt, October 31, 1916
My dear *Fräulein* Annamarie!
Bölcke is now no longer with us. It could not hit us pilots harder.

On Saturday afternoon we sat on high alert in our airfield huts. I had just started a game of chess with Bölcke – just after 4pm we were called to the front during an infantry attack. As usual, Bölcke led us himself. Very soon we came over Flers to attack several English planes – fast single-seaters who put up a brave fight.

In the wild dogfight that now followed, which only allowed us to shoot short bursts, we tried to force our opponents downwards by alternatively cutting them off, and we had done so often with success. Bölcke and I pursued just one Englishman between us when another opponent, who was being chased by our friend Richthofen, cut into our path. When we swerved at lightning speed to evade each other, Bölcke and I couldn't see each other for a moment, as our wings concealed our view – and that's when it happened.

How should I describe my feelings about the moment when Bölcke suddenly appeared a few meters to my right. He swerved his machine, jerked it upwards I mean, as we brushed against each other and both had to go to the ground! It was just a gentle touch, but at such high speed even that means a crash. Fate is usually so cruelly unreasonable in who he chooses: only one side of my undercarriage was torn away from me, the outermost part of the left wing from him.

After falling a few hundred meters, I regained control of my plane and was now able to follow Bölcke's, which I saw gently gliding, only hanging slightly askew, and heading towards our lines. Only in a cloud layer in the lower regions of the sky did his machine tip more steeply due to violent wind gusts, and I had to see how he could no longer straighten it out before landing, and how he hit the ground next to an artillery battery position.

People immediately rushed out of the battery dugout to help him. My attempts to land by my friend were not possible because of the shell holes and trenches. So I flew quickly to our airfield. They didn't tell me until the next day that I rolled over when I landed -- I wasn't conscious of it at all. I was very upset, but still had hope. However, when we got there in the car, the dead man was already being brought towards us. He died immediately at the moment of impact. Bölcke never wore a crash helmet and didn't buckle up in his Albatros either – otherwise he might have survived the crash, which wasn't too violent.

Everything seems so empty now. Only gradually do we fully realize what a void our Bölcke leaves, that with his loss the soul of the whole is missing. After all, he was our unique leader and master in every respect. He had a compelling influence on everyone who had anything to do with him, including his superiors, purely through his personality, and because of the naturalness of his being. He could go anywhere with us. We never had the feeling that something could go wrong when he was there, and almost everything went well. In this month and a half he has, with us, neutralized more than sixty enemy planes, and the superiority of the British is disappearing day by day. Now the rest of us need to make sure that his victorious spirit does not collapse in the squadron.

This afternoon was the transfer to Cambrai, from where his parents and brothers accompany their hero to his burial in the cemetery of honor in Dessau. His parents are magnificent people – despite all the pain, they're brave in the face of that which is unchangeable. That also gives me some consolation, but nothing can be taken away from my sadness over the loss of this extraordinary person.

Thank you very much for your last letter with the flower greeting. I was very happy about it, but I will have to wait some time before I can answer it – the experience of October 28 weighs too heavy on me.
    Loyally yours,
    Erwin Böhme

Lagnicourt, November 12, 1916
My esteemed friend!
That's what I have to call you today. That's because your letter, which knows how to find its way into the mood of my soul so sensitively, and wants to comfort and lift me up, provides a true service of friendship for me, for which I thank you from the bottom of my heart.

Outwardly, I've gotten myself back under control to some extent. But in the quiet hours I always remember the horrible moment when I had to see my master and friend fall next to me. The tormenting question arises again and again: why did he, the irreplaceable one, and not I, have to become the victim of this blind fate – because neither he nor I bore any blame for this catastrophe?

The days I spent flying, during which we could show that the spirit of the master had stayed with us, had the most healing effect on me. On the very day that his body was transferred to Cambrai, Richthofen and I took down a few planes. Another day brought five Englishmen. In total the squadron now has 74 victories. How happy Bölcke would have been!

It is a great relief for me that I now have my brother Gerhard close by. He's in a neighboring village with his field artillery regiment and he comes riding over to me almost every day. One afternoon he watched us knock three Englishmen down to the ground. He was so enthusiastic about it that he would have liked to switch to flying right away. But I won't tolerate that – two sons in aviation, that's enough for a lonely mother who has four sons at the front and who is already mourning the fifth, my sunny forest ranger brother,[63] who has been resting in Russia's soil since September 1915.

May you be well, my dear *Fräulein* Annamarie! Can I hear more soon? With warm gratitude for your good letter.
    Yours,
    Erwin Böhme

Hamburg, November 22, 1916
Dear *Herr* Böhme!
Today I just want to tell you about my life – maybe the entirely different images will distract you from your gloomy thoughts.

I feel very happy in my new job. It's quite a different thing when you're dealing with flesh-and-blood people in person than with the mere file numbers that they become in the central office. Through this personal exchange with those we are taking care of, I gain a deep insight into the lives and needs of our people.

Daily, three times a week in the morning and three times in the afternoon, I have to hold consultation hours in which the wives of our warriors express their wishes and submit applications. It is often quite colorful and sometimes a bit loud, since some women appear and make their demands in a rather brazen way.

The remaining hours of the day are used for home visits to the families of our front soldiers. This enables me to grow into my work; it makes it much easier if we already know the women who come to the consultation hours and their circumstances. During these home visits one sees how much poverty and crisis there is through no fault of one's own. It's endured without complaint and with quiet bravery, but we also see how there is so much despair as well as physical and moral depravity – things I have never dreamed of before!

While you were here we looked at some of the narrow passageways and courtyards in this area and found them to be very picturesque. But if you take a closer look and realize how the riffraff hide there next to those who are impoverished for reasons that aren't their fault, the joy found in those old buildings disappears all too quickly.

There is also no lack of elements who try to exploit, with shameless begging and fraudulent behavior, the boom in general willingness to help. That means: be careful! It happened to me the other day when a woman, who was by no means a particularly bad person, told me that she was going to take her own life, as well as the lives of her children, and staged a sort of suicide attempt in my presence.

You shouldn't let all this discourage you from enjoying your work – the grateful look with which many a woman who I already know greet me when they enter my office amply compensates for the ugly experiences that you can't avoid.

It is not always easy to make the right choice when faced with decisions. In addition, our hands are tied to a certain degree by the "guidelines" that are set up by those above us for approving grants.

There's enough money here. For the winter, the wives of soldiers receive an increase in state support: a woman with one child now receives 72 Marks per month, and with two children she gets 85 Marks. They can also get cheap meals from the war kitchen, a liter-portion for twenty Pfennig. For clothing and special emergencies, they also have to contact our "War Relief" office.[64]

Clothing is the most worrying thing. The supplies must be stretched as much as possible, so the review of all applications must be very thorough and strict. The coal shortage is already noticeable – hopefully the winter won't be too severe.

In addition to my full-time job, there are many other things that go on. Every third Sunday I have to look after the nursery in the Peoples' Home [*Volksheim*] in Barmbeck, together with my housemate Minni Cramer. There are about a hundred or more kids between ages three and twelve to look after and entertain while their parents listen to a lecture downstairs – last Sunday it was about Beethoven, with musical performances interspersed. We sing with the children, tell stories, solve riddles, etc. – I always have a lot of fun.

The newest thing that's been decided is to set up a boys' club in the Peoples' Home. Next Thursday I'll collect all the boys for the first meeting. In the winter they are supposed to be occupied mainly with building crafts– so I can teach them something, I've been taking handicrafts classes at the arts and crafts school for a few weeks now.

Each Tuesday evening, all Bermbecker employees gather at the home of Pastor Kiessling to get to know each other better and share their experiences.

In addition to the daughters of the house, we now have five young girls in the Sonnenau residence, all of whom have similar jobs. After our serious day's work, we are all happier at home. Last week we even had a party. That is, Rose von Wiese had Dr. Zahn and Pastor Kiessling, along with their wives and our patron, *Fräulein* Julia Kämmerer, come over for a picnic in the countryside. We transformed into a peasant family. I and two others were dressed like peasant boys. Small tablecloths, even pictures on the walls, were replaced by diplomas for breeding calves, pictures of horses, etc., as they did not genuinely match peasant style. It was a fun evening where we even feasted on various cakes that the kind guests had brought.

Incidentally, "Aspirin" was also invited, but she was unfortunately unable to make an appearance. I think you're getting the wrong idea about her after the few remarks that I wrote down so quickly, and so angrily. In reality, she's a highly respectable lady who works in an exemplary, self-sacrificing manner for the common good – what was probably strange to me about her is probably her distinctive Hamburg manner, which of course does not resonate with my uninhibited natural style.

Several evenings per week we attend the Sonnenäuler lectures in the lecture hall. Yesterday we all heard Brahms' requiem together in the beautiful, large St. Michael's Church – that serious music really goes beautifully with these great times.

Today the flags fly at half-mast in mourning for the old Emperor Franz Joseph.[65]

So there you have a picture of my current life.

I very much hope that you've now completely overcome the horrors of what you have lived through, and that the pain has been transfigured into silent sadness and loyalty, a harmonious requiem.

With heartfelt greetings, your
Annamarie B.

"*Jagdstaffel* Bölcke," December 12, 1916
Dear *Fräulein* Annamarie!
Pangs of conscience squeeze through the "Waterman's Ideal-Fountain" Pen in my hand. Really – I'm ashamed that I haven't thanked you yet for your splendid letter of November 22nd. I don't have an excuse, except that we're in

a state of hibernation in which we have sunk down into the almost non-stop foggy weather, which makes one a little bit dejected. You wouldn't believe how much pilots can sleep in winter! Anyone who comes to coffee before 9:30am on a rainy day generally makes themselves unpopular.

Actually, I shouldn't be telling you that at all, now that you're doing such a stressful, brave job. I was very interested to hear all those details about your work. It's not easy for me to imagine all the things you have to help coordinate. If all people were reasonable or at least had the good will to be so, it would be easy to get through; but since there are more of the other kinds of people, you always have to make concessions. I believe, however, that you have already found the right way of approaching your job: with your heart on the problems and your mind on the details!

We soldiers at the front can also raise some complaints about what the home front authorities sometimes give us from above. As long as war lasts, these two categories of "warriors" will always remain distinct.[66] Well, this problem will probably be the case for all time and will remain so until shortly before the end of the world.

All in all, everyone now feels, both at home and out here, that we are in a stronger position than we were a few months ago. One believes in Hindenburg's developing work, and so it must succeed…

Two hours ago, while writing the previous line, the AOK [*Armeeoberkommando* – Army High Command] broke the news that our *Kaiser* has made an offer of peace to our enemies. You can't guess anything about whether it will be successful. But even if it is accepted (which we all still believe here for the time being), Germany's fundamental approach to the world war will become even more unambiguous than before.

It's extremely funny to observe what kind of impression the word "near peace" makes on the individual. I must confess to you, as unbelievable as it may sound, that not one of us has let out a show of joy at the news. You just get used to the war. The possible upheaval to our "way of life" actually makes each individual think first about what effect this will have on their personal circumstances – that's how much people are creatures of habit.

First we have "Hedgehog" [*Igel*], Hans Imelmann (with one 'm', as he's not related to Immelmann), our spring chicken who will soon be 20 years old. Above all he thinks with quiet horror of the benches of his upper-level secondary school, from which he escaped with a leap of joy when the war broke out – now he's knocking down English ministers, which means he recently shot down his sixth plane, in which a former English agriculture minister happened to be taking part in an observation flight to the front, purely out of curiosity. Another thinks about the hardships of his forthcoming accountant exam. A third is an active guy from the boring Civil Peace Service [*Friedensdienst*]. A fourth made investments in Africa that have become worthless. Basically, however, everyone would be happy to put up with peace once the time comes

Since Bölcke's death we now have our second squadron commander, a Bavarian, *Oberleutnant* Walz, for whom we asked to be our leader. Unfortunately, we had to say goodbye on November 22nd to Kirmaier (also a Bavarian), who was given leadership of the squadron right after Bölcke's death.[67] Our English customers are a bit shy – we have to fly further and further away to get them. At one time there were five of us and we were attacked by two large squadrons at the same time – each of us had to deal with several opponents. I could still see Kirmaier chasing an already smoking Vickers two-seater,[68] but he had several chasing behind him – that kind of thing suited his Bavarian style.

I myself was attacked by a Morane monoplane at that moment. The silly fellow came at me from the front in a very clumsy, strange manner – now he lies near Longueval, where the Delville Forest used to be. That was my seventh "confirmed," because out of the three Russians, only one was credited to me, since the "ground verification," that is, confirmation by uninvolved observers, was missing.[69] They're very particular about it, but that's a good thing because it eliminates any attempts at boastful cheating.

Yesterday evening the new Air General, his Excellency von Hoeppner, was our guest.[70] On behalf of the highest authority, he said something flattering about our squadron, which was a welcome opening for us to make all kinds of requests that usually get stuck in official channels. For example, we asked for even faster machines, more cars, etc., so we could be more effective. The old gentleman was particularly pleased with the original furnishings of our mess: two large chandeliers made of captured English propellers and the like on the ceiling.

My brother Gerhard comes often and, fortunately, always has a horse brought for me. We then ride far through the region – it would be difficult for me to decide which is more beautiful, floating on horseback or in the Albatros. In the next few days we want to visit our oldest brother Erich, who is not far from Lille. Brother Martin is somewhere in the East in his giant plane. Mother's collection of postcards from her sons might look really colorful.

You may have wondered about the date of this letter. Yesterday the following decree of December 10 from War Minister von Stein arrived to us:
His majesty the *Kaiser* deemed it worthy to approve that the *Jagdstaffel*, which was led by *Hauptmann* Bölcke, who

died undefeated on October 28, 1916, was given the name "*Jagdstaffel* Bölcke". This gives us new incentive to prove ourselves worthy of the name of our master.

    I give you my most loyal greetings, my dear *Fräulein* Annamarie
        Yours, Erwin Böhme

P.S. I was very amused by your peasant evening. Is there a photo of it? Too bad that "Aspirin" wasn't there! From now on, I'll show my respect to the lady.

Jagdstaffel Bölcke, January 2, 1917
Hail and victory for the New Year, my dear *Fräulein* Annamarie!
This morning the post brought me an extremely aromatic little parcel: inside it was a Christmas apple of a perfection that usually only occurs in very old fairy tales. I'm not biting into it today, but tomorrow, knowing myself, the temptation will be stronger than any of my instincts to hold back. But I still have the silk ribbon to remember its past splendor – and the dear gift giver.

    But your letter gives me headaches and worry. Didn't my long letter with the Christmas and New Year wishes reach you? It should have arrived at the Hubertus Mill by the 25th at the latest. How lucky it is that I at least sent the telegram. But I'm bitter that you were at the parties without a letter of greeting from me. I think there was a lot of heartfelt and interesting things in it.

    It is a pleasure for me that you spent the festive season so happy and harmoniously with your family – on Christmas Eve my thoughts also went to those at the Hubertus Mill. Hopefully you've had a good time during the holiday season, because I don't think there's much left for the famous "Hamburg Cuisine" anymore.[71]

    We also celebrated Christmas Eve quite nicely here, in the way that a warrior can. But that's just a Christmas "substitute"(*Ersatz*).[72] Proper German Christmas can only be celebrated in intimate family circles. I also just withdrew as soon as possible and entrusted myself to my beloved violin.

    We had all sorts of things in the air to do just around Christmas time. On the 26th – and this is highly inappropriate in light of the "Festival of Love" – I shot down my number 8, over by Courcelette, just as he emerged from his cloud and was about to do something to one of my comrades.

    I can imagine that reading Bölcke's "Field Reports"[73] will fascinate you to a large extent, and it will have a special appeal for you since understand and know some of the details better than other people do. For example, you'll know the mysterious "*Leutnant* B.", who is mentioned several times.

Bölcke's father immediately sent me one of the first copies. I admire the old gentleman who, in his reignited pain, was able to put this book together as a memorial to his son.
Frequent fog – the really thick, mushy kind! – puts a heavy strain on my well-being and mood. It would probably be good if I relaxed soon. But I can't make up my mind about whether I should leave the squadron, because if there are bright days in between, I don't want to be absent.
To make things worse, I also had to take over the business of our administrative clerk, who had gone away for a few days – but I'm quite good at doing this on my own. All this office work even starts to deflate what usually gives me the greatest pleasure: writing a long letter to *Fräulein* Annamarie.

    Believe anyway in the friendly mood of your ever faithful,
        Erwin Böhme

Partenkirchen, January 16, 1917
Greetings, *Fräulein* Annamarie![74]
Just took out a subscription for fourteen days of winter sun here. I was simply sent on vacation for three weeks, "in need of relaxation," and I can already see that it doesn't do any damage. First I was with my mother for two days. I've been here since yesterday.

    You know how I love the Swiss mountains, especially in winter. You're not allowed to go there now. I wasn't familiar with the Bavarian Alps yet – so I flew to Partenkirchen.

    It seems that I hit the right spot here. The only thing that bothers me are the many uniforms. It's filled with officers on leave and in need of a rest. Yesterday on my first walk I met several friends and finally I ran into one not overly sympathetic captain from the flying profession in the army, who immediately wanted to hog my time.

    That's just no good for me. I don't want to hear anything about war and war talk here, I just want to be human again. So I avoid the Wigger's House,[75] where I actually wanted to live, but which is full of comrades, and I avoid strolling between Partenkirchen and Garmisch. If possible, I'll retire to solitude and want to brush up on my old winter sports skills – at the moment the only thing missing is the necessary snow.

    Today I took a wonderful walk to the small Badersee [Bathing Lake] and back. Lord God, how beautiful the world is! And how nice it is to breathe peace again after so much war and murder!

    On my lonely hike, it was very strange, but heretical thoughts came to me. When I'm out on the field I think of nothing but war and my obvious duty to harm the enemy as much as possible. But here, removed from the war, in this divine peace of the mountains, the thought came to me: isn't this mutual, endless slaughter actually madness?

There are also many people of value among our enemies whose life's work, possibly irreplaceable, is destroyed by a stupid accidental hit! What chivalrous opponents I have found among the English with whom we fight. What decent fellows among those we captured! But I have to shoot them like they're wild game.

But above all: what kinds of consequences does this murderous war have for our people as a race![76] It's like this: the best and fittest are out there and give up their lives – but the scum at home, the slackers and unfit, stay alive, propagate and degenerate the people and our national character.[77]

Once I'm back out there, I'll probably not understand this train of thought myself. But today it came to me, and so I shall entrust it to you.

Will I still find my squadron on the Somme when I return to the front? So much is being reorganized now. My friend, Richthofen, just received the Pour le Mérite; after sixteen victories. Two days before my departure I quickly shot down number 9 – yes! I don't want to think about the war now.

I will stay here for about two weeks, then I'll greet Bölcke's parents in Dessau and then carry out an absolutely necessary official inspection of the Bermbeck war relief office. Understood?[78]

With these hopes I greet you with a happy "hail to the mountains" [*Bergheil*]!
Yours, Erwin Böhme

Partenkirchen, January 28, 1917
Thank you very much for your good wishes on my vacation days and for the wise advice. I didn't even know that you're familiar with and love Partenkirchen – since learning that, I've looked at it with completely different eyes and think about how on each path *Fräulein* Annamarie may have cheered at the sight of all the beautiful things that surround one here.

I'm very satisfied with my accommodations. However, the snow situation was not very favorable for skiing – there was still not enough and the little there was is too powdery. So I mainly went on long hikes. That was nice too, and refreshing in the clear, cold weather.

This winter has real spirit and is to my liking. Despite the coal shortage! One should close all the schools – for the good of young people, and above all they should close the writers' retreats – for the good of mankind!

With cunning and skill, I've pretty much been able to keep my solitude and evade the warriors and their war talk. However, as a result, I was not able to convey the greetings from your roommate, Minni Cramer, to her cousin Plüschow. Plüschow lives near the Wigger's House, which is now the officer's mess – ten horses couldn't drag me in there – and I never met him on a walk. But I read his "Flieger von Tsingtau" while here.[79]

One can elude the warriors, but war cannot flee from the war even in Partenkirchen. Today I received the sad news that our Imelmann, the youngest of our squadron, died on the 23rd. He was a comrade without fear or reproach. At the same time, I learned that the squadron is suffering another serious loss. Richthofen is leaving because he has been appointed the leader of *Staffel* 11 in Douai. I'm pleased, because Richthofen is undoubtedly destined to play a major role in aviation. But he leaves a big void in our squadron.

Influenced by this news, I made the decision to return to my squadron as soon as possible before the end of my leave. I'm needed there. This decision demands a very heavy sacrifice from me: I have to miss my visit to Hamburg. How I was looking forward to it! It should have been the high point of my leave. But iron duty comes first.

So I'll just say goodbye to my little mother and good aunt Lossow in Holzminden and then hurry straight to the front

When will I see you again, my dear *Fräulein* Annamarie?!

I greet you from my heart,
Erwin Böhme

Cambrai, February 13, 1917
Dear *Fräulein* Annamarie!
Even at my second-rate school I mostly got a 3 or 4 in calligraphy[80] – but today you'll probably grade me even worse. That's because I'm writing in bed and don't have a left hand to hold the paper. The bed in which I'm writing is in a field hospital [*Kriegslazarett*] in Cambrai. And I find myself in this field hospital because the day before yesterday some malicious Englishman, who by all rights is no longer alive, treacherously shot me in the left arm. It was a Sopwith two-seater,[81] which I had already pummeled so that he was going down, and so, in a surge of hunters' nobility, I spared him – that's what I get for my nobility!

Don't fear for my life. Not even for my arm! The bone and nerves are unharmed and the bullet is still in there. The fact that this hurts so damn much is just a minor matter – I'm only angry that I have to let down my squadron, especially now that the spring-time anger has started.

So you don't accuse me again of reporting one kill in every letter, I'll report three this time: I got two on my first take-off after leave on February 4th, the next one on the 10th, and this treacherous one on the 11th should have been number thirteen – this unlucky number seems to have had an impact after all.

The gods only know how long I have to stay here in the hospital and whether I can go to a hospital at home to fully recover.

Please send some greetings soon to cheer up the hospital inmate who is condemned to loneliness and inactivity. My address for the time being: *Jagdstaffel* Bölcke, *Feldpost* 323.
Loyally yours,
Erwin Böhme

Cambrai, February 29, 1917
Dear *Fräulein* Annamarie!
Your good letters were a salvation from my solitary confinement. But the fact that you shared your valuable "triumph",[82] which you luckily acquired, with me so amicably made me really annoyed at first, because you need such sustenance much more than I do with your work and insufficient nutrition – but then I felt how nice and loyal you are and it was a ray of light in my hospital mood.

Today I can write to you again from my desk. After being confined for fourteen days, I was released from bed and house confinement on Sunday. Coincidentally, my brothers Erich and Gerhard visited me on that day – so I could attempt my first little walk in the sunshine with them. Since then I've walked a little bit from day to day, and today I've even had coffee with a fighter squadron located near Cambrai.

My wounds are beginning to heal. I now have to do certain stretching exercises so that the elbow joint can be used again. Incidentally, this matter will probably have some after-effects in the years following the "outbreak of peace," since most of my fur jacket was applied to bind my wound and the uniform office will undoubtedly object to such an improper use of an official item. This will be a treat for the red tape heroes![83] – imagine the whole sheafs of files that can be filled with something like that.

It's not clear yet whether I should go to Germany for follow-up treatment. I wouldn't have much patience for it now, because we're totally swamped and one needs to be useful here. Prince Friedrich Karl, the great sportsman, is joining us as a new member of the club.

I'm out of letter writing paper, and out of thoughts. You wouldn't believe how stultifying hospital existence is! Don't let your loved ones go through that!
Erwin Böhme

Jagdstaffel Bölcke, March 17, 1917
Dear *Fräulein* Annamarie!
Yesterday I finally escaped the claws of the doctor with a boost from a friend. Today I even flew a lap of honor and relief over our new airfield. But it was only a modest flight.

My arm is still a bit stiff and can't be fully extended.

As a result, I still have to go home, namely to Düsseldorf, for some follow-up treatment (orthopedic exercises). I asked whether or not they'd like to send me to Hamburg, but they said they didn't know a proper orthopedist (nice German word!) there.[84] But I also like going to Düsseldorf, because that's where I'll find my dear brother-in-law, Kohlschein.

You ask what we think about the overall situation and the prospects for peace here? We don't think about it at all! You do what is demanded in the moment and leave the rest to God and Hindenburg.[85]

Incidentally, because of the transparent newspaper people, you at home always know more than we do, and more than what is true.[86]

Tomorrow we're off to the Rhine. I secretly hope that the remedial gymnastics will make me so strong that I can finally risk a bold leap to Hamburg.
Sincerely yours,
Erwin Böhme

P.S. Last Sunday our General of the Air Service surprised me with the telegram that the *Kaiser* had awarded me the Order of the House of Hohenzollern with Swords. So I've been walking around with this new jewelry for almost a week now. Now both of us, the loyal letter writers, will soon have to have our picture taken together!

Düsseldorf, April 4, 1917
Dear *Fräulein* Annamarie!
I'm now sitting at my desk with a very dark frown. Instead of the planned onward journey, I have to plan for a return journey to the front tomorrow, and I had thought of such a nice agenda for the week that had been added to my leave. The telegram just arrived: my additional week has been cancelled and, on top of that, I've been told that I'm assigned to the Valenciennes fighter squadron school – they probably think that I'm on crutches!

Well, that's all been finalized – I sent a telegram to protest against the dreadful school-teaching. But I'm angry, very angry that my wonderful plan to spend the holidays with you at the Hubertus Mill has been so cruelly cut short. At the front they seem to have forgotten that tomorrow is Maundy Thursday and in four days it's Easter.

Well, that's not helpful – you just have to clench your teeth.

I send my warmest Easter greeting to you and everyone at the dear Hubertus Mill
From yours, Erwin Böhme

Jagdstaffel Bölcke, April 8, 1917
My dear *Fräulein* Annamarie!
Today I wanted to help you look for Easter eggs at the Hubertus Mill. Now I want to at least visit you from afar with a letter.

This time I made my return trip in a somewhat gloomy mood because I always had to think about the Easter celebrations that I didn't allow myself to have – the return was thus even sadder. When I arrived late in the evening in Cambrai, the first thing I learned was that two good comrades had fallen: König and Wortmann. There were the last two men in our squadron who were from the original Bölcke unit. Now only Richthofen and I are still alive.

The chalice of Valenciennes seems to be passing me by – I even have a chance of getting to lead *Staffel* 5.

By the way, you mustn't think of Valenciennes as something in which being given a command there meant being pushed aside, like being given a district command in peacetime. On the contrary – it is a sign of great trust. Valenciennes is our only fighter squadron school, through which all of the many young fliers who are not enlisting in fighter school have to go through to be trained and weeded out. It's therefore a task that is important and requires responsibility, because the quality of the whole replacement of the fighter squadrons depends on it. It's just – I don't like this task. I'm such a weird old codger[87] that I prefer the smallest dogfight to an entire morning of school work.

Here we've got a terrific air operation going now. The English are coming in mighty swarms with new, very fast airplanes. But now, at least in most squadrons, we've got a level of passion that's really magnificent. What a sense of joy Bölcke would have!

This morning I was with Richthofen, who has now become captain. He had just brought down number 38. It's amazing to see the high level he's reached with his squadron in such a short amount of time. He's got a lot of great guys around him who will go through fire for him; his younger brother Lothar has also recently joined the squadron. Richthofen himself is in the best of health. Even though he takes off five times on some days, you don't notice a trace of fatigue. What pleases me is that he is so completely unpretentious, a top-notch, but completely natural person – he is always particularly kind towards me. It would be good if he were soon placed at the head of all fighter aviation. According to Bölcke, he would be the man to do it.

At night, Richthofen and his men organized some cheerful target shooting because the English wouldn't let him sleep peacefully. After there was some quite low-level night-flying in which bombs were thrown at his airfield a few times, he had 17 English machine guns that had been captured in aerial victories installed near the mess. Thus they waited for some guests and when these guests boldly descended to fifty meters, they were surprised with blazing fire. In this way they also shot down three in one night from the ground – something new for a pilot! Now they'll probably be safe from nocturnal disturbances of the peace for a while.

You'll probably not find much Easter spirit in the letter. You see, I'm back in the middle of a war. I hope it stays that way and I don't have to go to Valenciennes.

From the heart, your Erwin Böhme

Valenciennes, April 25, 1917
Dear *Fräulein* Annamarie!
So, as I told you briefly, the *Reich* did not take a pass on me, despite all my foot-dragging. So I'm staying here in a castle near Valenciennes in the middle of an old park and using my experience to train young fighter pilot apprentices. That takes a lot of work, because we have eighty little elves here.

It's very comforting to me that my "school director," Hauptmann Zander, whom I already know as a brave comrade from *Jagdstaffel* 1, is a competent sportsman and all-around splendid fellow. Incidentally, his civil occupation is based in Düsseldorf. Höhndorf is the next one to be sent home. Höhndorf will soon be coming here as a teacher – then we'll be able to do a free-lance flight to the front from time to time. A summer bathing establishment has already been set up in a nearby river. Unfortunately, we have no use for a row boat like the Bölcke squadron now has on the canal.

As you can see, I've come to terms with my situation to some extent. Only the meaning of the whole enterprise is not so completely clear to me. It used to be that we who were first formed into the hunting squadrons last fall simply switched from the two-seaters to the single-seaters and started shooting – and look: it worked. But today you have to "retrain" men rigorously. I mean: either someone has what it takes to be a competent fighter pilot, and then it can be done without "retraining," or he doesn't have it. Then "school" can't teach him either.

But that may be a perspective that was true back in the day, when only people with passion applied to the new job of fighter aviation, but it's no longer quite true with the current mass influx of applicants. In any case, a sharp and ruthless selection must now take place, and that seems to be the most important part of my task. Of course, this task of "weeding out" is not exactly pleasing for us poor "college teachers."

When I've finished my schoolmaster's work in the evening, I hear on the telephone what Richthofen and his squires have achieved during the day, and I get spiritually intoxicated by their successes. I also fly over to Richthofen's as soon as possible – one always wants to visit *Staffel* 11; the same spirit and the same passion reigns there as in Bölcke's best times.

It is delightful to see Richthofen with his "Moritz".[88] That's his close friend [*Intimus*], but a four-legged one. He once bought him as a cute lap dog somewhere in Belgium for five marks. Moritz, inseparable from his master, thus took part in the world war. He was also with us in Russia, and has meanwhile become a huge beast who doubtlessly bears somewhat of a resemblance to a Great Dane.[89] However, his master insists strongly that Moritz has a very pure pedigree. Richthofen loves his Moritz dearly, and as a result he also very much loves my Brother Gerhard, who recently cured the animal of something for him. Moritz isn't exactly to my taste, but you shouldn't let it be known if you don't want to offend the good Richthofen.

My preference is the German hunting dog, more specifically the Dachshund, the eccentric humorist and opinionated character. The worst for me is the St. Bernhard, whom I know well enough from Switzerland. If he is removed from his element, the mountains, and placed in the lowlands, he becomes a boring, grumpy, asthmatic patron – he always seems to me to be like some fat, exhausted, unapproachable banker in a lounge chair.

By the way, my brother Gerhard massaged my arm again. Indeed, I've always said: whoever has something wrong is an ass if he doesn't turn to a capable veterinarian right away.

But first, let me thank you for your lovely "spring greetings"! It really seemed as if it finally brought the proper spring with it. For a few days the breeze was milder, and in our park there appeared all at once some anemones, snowdrops, primroses, and daffodils, all of which we'd been waiting on for a long time – some of them are probably appearing in bloom for the first time in this meadow. But today we've got to turn the heat up again – I can't deny that this winter is gradually becoming too intense even for me. There will probably be no spring at all this year…

I just got interrupted. Richthofen came to me (he'd quickly dealt with number 47 on the way) and told me that I should stand in for him during his forthcoming leave that would last several weeks. He is at a loss as to who to transfer leadership to because the best in his squadron are away: Voss, who already has 24 wins and the Pour le Mérite, is on leave himself, and Schäfer, who has crossed off 23 opponents in less than two months, has just been transferred to another squadron. Richthofen cannot wait for Voss to return because he is supposed to be on leave as soon as possible – they want to bring him to safety; it's very similar to when Bölcke was sent to Turkey.

You can imagine how tempting Richthofen's request is for me. What could be more wonderful than being able to lead this elite squadron for a few weeks?! But I guess nothing will come of it – they won't let me leave here.

My brother Martin writes enthusiastically about his bombing missions in Russia. But I have to sit on the sidelines and can only send my greetings as,
         Your poor schoolmaster

Hamburg, April 29, 1917
Dear *Herr* Böhme!
Today I wanted to write a very long letter to you – but it doesn't seem to be that much.

Do you know where I'm writing to you from? I'll say it without blushing: from the bath! It's because your beloved winter is so intense – and we're just two days before May 1st – that early this morning my fingers were all frozen. We haven't had coal for central heating for a long time. So we heated up the bath stove and now the five of us girlfriends are sitting, squeezed tightly together, in this little space. One of us wants to read Wagner's letter to Mathilde Wesendonck,[90] the second wants to do her accounting, and we three others want to write letters. Since only small tables could fit in our little cage, we have created further writing opportunities in our bathtub using a flat board and the top of a box crate.

"Such a cheerful prison!", you might think. Yes! But you can also imagine my inner concentration as I sit over my letterhead -- especially after the girls found out that I want to write to you! Now they keep interrupting me and dictating all sorts of nonsense to write to you. By the way, ever since the others found out that I was getting letters from a "full-fledged" (a popular word here right now) fighter pilot from the front lines, my standing among the girls has gone up significantly. Only Minni Cramer can compete, as she has her cousin, Plüschow.

Now I could tell you about how the ever scarcer stocks of goods are causing me a lot of concern as I think about those whom I have to protect. I could write about how the merry band of boys I'm looking after in the boys' home is having a lot of fun and we've now dug up the whole garden, which was actually supposed to be their playground, to grow potatoes there. I could also write about the colorful cabbages that we're supposed to be planting in the gardens instead of flowers. I could describe how we're no longer going to have chocolate because the Hamburg war nutrition office in Reichardt wants to ban the sale of it here, so that only the people of Wandsbek will have the pleasure of chocolate and you have to set up some "connections" there, or maybe you'll want to capture some on your morning flights at the front.

I could write to you about all this and many other things – but I can't do that any longer. This is because our industrious group of bathing colleagues has become a high-spirited morning brunch party.[91]

Gertrud Overbeck, our jolly Rhinelander, got a crate of wine from home (I'm writing on the lid of the crate now), and she just got a good bottle out of the crate, so that while we warm ourselves up on the outside we don't miss the chance to warm up on the inside too.

And so we grab our glasses, clink them together, and shout enthusiastically – hail and victory to our brave German combat fliers!

Minni Cr…Rose von W…Gertrud O…Thea P…Annamarie B…

Valenciennes, May 9, 1917
Dear *Fräulein* Annamarie!
Thank you and your lovely drinking buddies for the cheerful Sunday brunch party greeting, which I really enjoyed! I appreciate such a quiet brunch on a Sunday morning after a good night's sleep. Of course, if you didn't have an uncle who owns a winery on the Rhine, you'll probably have to hold back a bit more. Well, that's just a minor evil – more a problem for the rowdy types whose life revolves around being regulars at the pub.[92]

Of course, nothing came of the swap at Richthofen's squadron. They don't want change to cause disruption here at the flying school – and they're right about that, because the issue of timely reinforcements to meet the great demand from squadrons is important now. By the way, they suspected upstairs that I myself contrived to sneak out of the school this way – although this time I'm really completely innocent and was only responding to Richthofen's wish.

Today, Richthofen's younger brother Lothar, who is leading the squadron for him right now, was with us for lunch. He wanted to get some information from me about the youngsters and hire new talent. The day before yesterday, for his number 20, he shot down Captain Ball, the English "Bölcke," after an extremely hard fight. He was, we believe, the leader of the "anti-Richthofen squadron," newly formed by the dashing Royal Flying Corps.[93] Our pilots have often had to deal with this in the last few days and have already shot down several of them – so the English have not yet experienced much joy over the price that's been put on Richthofen's head.

He is currently shooting Roebuck in Silesia. Before that he got a wood grouse in the Black Forest. Soon he'll stalk some game that won't even exist in Germany anymore: the aurochs or bison. Prince Pleß invited him there, right about the same time he was ordered to the "main headquarters," on his 25th birthday, on which day the emperor and empress sent their congratulations to him.

Hunting is Richthofen's whole passion and for him it's the priority. And he has his hunter's eye to thank for his fabulous successes in the first place. With an eagle's sight he sees his opponent's weakness and, like a bird of prey, he pounces on his victim, who inescapably succumbs to him. I don't think he has much use for flying itself. He's probably never done a loop out of sheer love for the sport, or for the fun of it, and he's strictly forbidden his squadron from doing "acrobatic acts," as he calls them.

For my part, I feel pretty much like an old wreck now.[94] The rest of my resilience is shot to hell, and for the first time I feel like I got "nerves" during the war.[95] Where this comes from, I don't know. Maybe the long, stupid life in the field hospital is to blame. Maybe the tired old school system is to blame – in both cases it would be a natural relaxation compared to the sudden relaxation that hit me in the middle of the turmoil of the war. If I were at the head of the Richthofen squadron all the time, I would probably be a completely different, fresh and energetic guy.

Maybe what also contributes to my gloomy mood is that I now have too much time to think about how and in what form my future life should develop. In these bottomless times, with the current uncertainty that surrounds everything, that's a difficult topic.

Recently, the sister of my squadron mate Philipps, who was married in Bielefeld and fell in October, sent me a copy of two letters from her father from Africa, who is now working in Wilhemsthal, very close to Neu-Hornow, as a doctor for the Germans who were interned there. They are doing relatively well since being interned. Their treatment by the English seems to be considerate. One can therefore hope that our facilities in Neu-Hornow have also been preserved; it even seems that the English have put them to use.

But that's all uncertain, and besides, I must confess to you that my desire to return to Africa in the long-term is not all that great.

I would love to be a farmer – on my own plot of land, of course. That's a thought of mine that didn't just grow out of the current potato and bacon shortage, but it has been seriously preoccupying me for a long time. To live as a baron [*Freiherr*] on your own land, in closer connection with nature, far away from all the disgusting big-city hustle and bustle, that seems to me to be happiness worth striving for. I'm thinking of Kurland or something.[96]

The best thing about Valenciennes is that the English sometimes come here too – there's always something to do. Unfortunately, they always fly so high that you can't catch them by the time you reach the front. On one such occasion the night before last, I was just about to see Voss opening his latest guest performance at the front by shooting down a fat Vickers[97] – he had just returned from leave half an hour earlier.

…Just got back from a three-hour writing break! Just as I was writing earlier on this topic, another English squadron with bombs appeared over Valenciennes. When I had luckily caught up with five of them at the front, they dove into the clouds and I had to pull away again. One the return flight, I ended up landing at my old squadron airfield.

After this happy intermezzo I have to go right back to another hour of teaching school and therefore must close quickly.
    Heartfelt spring greetings!
    Yours, Erwin Böhme

Hamburg, May 17, 1917
Dear *Herr* Böhme!
Although it is actually a sin to sit in the living room on such a spring morning, I want to write you my Pentecost letter on this radiantly beautiful Ascension Day. But during the week my office doesn't give me time to write letters. Next Sunday the group from Sonnenau wants to go under the beech trees in the Saxon Forest, and anyway I've already enjoyed my happiness this Saturday and Sunday and heard the first nightingale and cuckoos call.

I was in Ütersen again with my dear aunts. I like going there. It's such a cozy, peaceful town, and for me it's a piece of home. My mother spent part of her youth there, and as a child I often went there on vacation – it's full of old friends and acquaintances.

On Sunday morning we visited two old ladies who befriended my aunts. They're nieces of Field Marshal Moltke, whose sister was married to Provost Broecker in Ütersen. Of course, the conversation soon turned to the war and aviation, and since I had taken my aerial photographs from Hamburg to show my aunts, I brought them out there too. When the ladies came to the picture of you standing in profile in front of your terrifying dragon in Kovel, they cried out with one voice: "No – such a resemblance to Uncle Helmuth!" Then they ran and got all their photographs of the Field Marshal from their tables and chests of drawers. In fact, your resemblance to Moltke's pictures from his younger days is striking. I congratulate you, *Herr* Böhme, on your fabulous military career, which, given this similarity, is undoubtedly still ahead of you!

In the afternoon the aunts took me for another walk through the marsh. We returned to a farmhouse that they knew about and were served – without ration cards – delicious grass-fed butter and fresh milk.[98] The portly farmer wore an iron cross first class, but didn't want to say what he got it for. He just said dryly: [in dialect] "They gave these to everybody." But soon there unfolded a lively conversation about the fighting at "Verduhn" and on the "Eisne" and about our prospects for the war.[99] I really like listening to the beautiful, genuine Low German [*Plattdeutsch*] dialect, but unfortunately I can't speak it – my aunts in Ütersen understand it all the better. One feels so safe and at ease with these Low German peasants. There is such a primal force and yet also considerable sensitivity and sound judgment in these thick peasant skulls.[100]

As we hiked back towards Ütersen along the meadow paths and over the ditches with many yellow marsh marigolds, a heavy storm cloud hung over the fields, lightning bolts hit the ground in mighty spikes, and it was a wonderful picture when four storks flew by the black cloud covered forest, on their way to their nests. Before the rain began to pour, we were just able to escape to the first farmhouse that we could find, where we were immediately asked to come into the parlor. Here we had a poignant experience. The farmer's son was still in his field-gray coat, but almost blind—he carefully tapped around the house with his cane. Suddenly the door opened and, led by his wife, in came a fellow of his age, also in a field-gray coat and completely blind. The way in which the hands of the two old comrades greeted each other for a long time – that was an image that I will never forget. I have never so profoundly seen the horror of war. It was only when we walked home through the rain-fresh meadows in the glow of the evening sun that my inner shock gradually dissolved…
    Yours,
    Annamarie B…

Valenciennes, May 23, 1917
Dear *Fräulein* Annamarie!
Many thanks for your Pentecost letter, which I value particularly highly as a sacrifice that you made for me. Yes, you are right: it's a shame to have to write letters in springtime – at most, one should write poetry on such wondrous days and nights! But that's something you have to learn.

Now we live outside all day long, unless there are classes. After a number of warm thunderstorms, nature is now showing a dramatic blossoming of oranges. It's a fruit that's beyond compare! Our large park, with its radiant freshness of spring, and with the old and mysterious animals that it hides, is a source of inexhaustible pleasure.

I find particular joy in the countless armies of birds nesting in the old trees and the impenetrable bushes of the somewhat overgrown park. Nightingales, orioles, warblers, robins – all my old friends are there. There is jubilation and warbling and sobbing all around our castle from the early morning until late at night. There are also several pairs of magnificent wild doves. So I always have to be ready to prohibit any of my scholars from becoming game hunters–they will soon find enough opportunities for other aerial hunts!

Another great accessory to this fairy tale image is a very funny and aggressive donkey – he always comes to our mess hall to get leftovers of bread.

So the whole area is idyllic! It's just that one should be able to enjoy all this without the war – May is a bad time for human conflict.

Isn't it true that life in the countryside is the only good thing for a sensitive mind? At least, I read this into your experiences on the rural walkabout. There's much that townspeople, those poor unsuspecting souls, would find annoying or uncomfortable: thunderstorms, wet meadows, people speaking Low German, etc. – all of this belongs to the realm of nature, and street cars, newspapers, even the cinema would rather have nothing to do with it. Without wanting to belittle the value of our civilized cultural achievements, I believe that once you have become estranged from nature and no longer recognize the easy-going charm of rural life, you are no longer able to find any real joy in what is really valuable in the arts. You can see this in the cultural taste of average people from Berlin.

A peasant, especially one from northern Germany, is not in the habit of sharing his feelings, but then one is all the more surprised at his incredibly strong sense of judgment, as long as things are not too complicated.

So you should go often for the grass-fed butter and fresh milk, even if the conversation in low German is still difficult! In a pinch, Fritz Reuter, who should be read from time to time, can be helpful.[101] My brother Erich always carries his novel *Farming Days* [*Stromtid*][102] around with him and has not parted with the book even during the war. When we were boys, we used to speak in "Reuter-talk" to each other when we wanted to be secretive.

I've read some good stuff by Hermann Löns lately.[103] They all take place in the grasslands and marshes; in his descriptions of people I'm sometimes reminded of Frenssen.[104] To me they are to be treasured for their unsurpassable depictions of animal life. Unfortunately, Löns died right at the beginning of the war – for him, given his circumstances, death was probably a form of liberation.[105]

Richthofen is still in Silesia. His brother Lothar has now also packed up: three days after he'd recently visited me, immediately after he shot down his number 24, he received a shot in the pelvis from ground fire – but he's out of danger. The squadron keeps going on bravely without him, especially Voss – he's quite a fabulous guy! Only I've got to sit on the sidelines and watch.

This letter must go if it's to get to you on Saturday. For Pentecost I wish you nightingale songs, grass-fed butter, pure relaxation in Ütersen with your good aunts, Reichhardt chocolate and above all a happy heart!
Faithfully yours,
Erwin Böhme

P.S. Now it's about a year ago when three pirates of the air crashed the Hubertus Mill!!!

Valenciennes, June 10, 1917
In the unlikely event that your honorable excellency should still remember the undersigned -- this is how I should I really begin my letter. But, dear *Fräulein* Annamarie, if you don't want to make any internal or external progress, I should only discuss: school, walking around the park, and watching what other people can do. But you don't want me to repeat that same old song and suffering over and over again.

In the last few days, however, I finally experienced something again, and you should hear about it right now.

Yesterday evening I returned from a three-day fact-finding trip to the front with my local commander, Zander. We visited a number of fighter squadrons from Lille on northwards to Zeebrugge – naturally by plane. We just witnessed the great English offensive in the Wytschaete salient.[106]

It was a powerful air operation, and we also got shot up a few times. Right after the first skirmish, I was shot in the front cylinder and had to slip on home (in this case, to Menen), softly crying. Yesterday, I came across a Nieuport of the latest design and had already pushed it down from 4,000 meters to 500 meters when he finally noticed that I had jammed both guns, and he escaped. Yes, my "thirteenth" seems to be quite difficult to get.[107]

By the way, this short trip was quite a splendid change for us two poor reserve-lines hogs [*Etappenschweine*] (please don't be repulsed by this poetic expression – that's the name of all the "warriors" who perform their activities behind the front, whether out of innate bravery or because of official decrees).[108]

Zeebrugge was especially interesting to me. What I got to see there was all new to me – seaplanes of various kinds, torpedo boats, etc. I also took part in a submarine journey, but only over the surface, from the mole to the locks. The boat had sunk 47,000 tons of shipping on its last voyage on the Franco-Spanish coast, as well as an English destroyer. With the commander, *Kapitänleutnant* Wassner, we then sipped a very pleasant pineapple punch (made from tons of captured goods!) and sang the most beautiful U-boat songs –

it was very festive.

Incidentally, it was the first time since the outbreak of the war that I saw the sea again. From the air, this smooth, linear Flanders coast offers a very strange picture.

You've probably read that Schäfer has fallen – he was an unusually capable guy, but he's been particularly vulnerable since taking over a young, untested squadron. The day before I'd eaten a piece of shoot-down cake[109] with him in the garden to celebrate his 30th victory.

From a purely animal point of view, I'm doing quite well – with daily sunbathing and swimming pools, thick milk, strawberries, and soon we'll have cherries, no kidding! Unfortunately, none of these are remedies for an ailing heart that thirsts in vain for action. I've only ever heard that overwork and excitement can make people nervous, but now I know that it can also be caused by lack of proper activity and tediousness. But I've already reached the safe stage of resignation – I've become so docile and patient that I'm always on the lookout in our park to see whether or not a dove of peace wants to reveal himself among my wild pigeons. What's a war to me if I'm not allowed to take part in it?! I believe that I would accept it without any pain if peace suddenly broke out. I'd just for the life of me like to first smack down one of the newly imported American planes.[110]

With this hope I send my faithful greetings,
Yours, Erwin Böhme

Valenciennes, July 3, 1917
Hooray! Now a new life begins! Yesterday I took over *Staffel* 29 from Wolff.[111] My joy is enormous – I thought they'd already put me in the old-age home. Now I want to show them what I can do.

And it's also magnificent that my dear *Fräulein* Annamarie has also become a "squadron leader" and has been in charge of the Barmbeck squadron of the Hamburg War Relief office for the last three days. It's great that they've recognized you as someone to be entrusted with this office, which was previously managed by a Hamburg lady! Now be aware of your increase in power and be sure to always be sufficiently strict against your subordinate members – I hope "Aspirin" has set an example in how to do this.

My promotion is related to the fact that a large "fighter wing" ["*Jagdgeschwader*"] is now being formed under Richthofen, consisting of the four squadrons, 4, 6, 10, and 11.[112] This is another important moment in the history of aviation, and it has been our wish for a long time and it's a necessity that will enable us to oppose the swarms of locusts with unified, coordinated large formations. Since Richthofen, as leader of *Jagdgeschwader* 1, is to a certain extent a commander over four squadrons, he can of course no longer lead his previous *Staffel* 11 as chief, but Wolff has taken over this spot, and I moved into *Staffel* 29 in Wolff's place.

Soon I will have the joy of seeing my brother-in-law, Kohlschein, who wants to paint Richthofen, here on the Western front. So when I was with Richthofen a few days ago, I found him very much pre-occupied with arranging the fighter squadron and, moreover, annoyed at the many newspaper reporters and cameramen who are always besieging him now. So he made a slightly sour face at first – but when I told him that Moritz would of course be included in the portrait, he was of course all for it.

I'm writing these lines while sitting on my packed suitcase. Tomorrow we go to my squadron; it's located in Bersée, between Lille and Douai.

Greetings to the newest district leader from the newest squadron leader.

Jagdstaffel 29, July 16, 1917
Dear *Fräulein* Annamarie!
"Whatever you wish for in Valenciennes, you get in abundance in Bersée" – namely work, real work, which makes the heart happy.

I have plenty to do here, as most of my squadron is composed of young fighter pilots. But there are very capable people here. Admittedly, with the exception of a few elite squadrons, the overall spirit in the squadrons is no longer the same as the beautiful times from a year ago when the first fighter squadrons were made up exclusively of passionate people who knew each other well and never let themselves down in any situation. But that's not surprising considering the mass replacements that many new fighter squadrons now require. Single-seater aviation is a field where aptitude plays a far greater role than in any other branches of aviation.

Above all, I have to tell you that the difficult and tiresome number thirteen is finally behind me. It was an English Nieuport single-seater that attacked me at 5,000 meters – I then knocked him down almost right over our airfield. For two others that I shot down quite far over there I couldn't get accurate confirmation from ground observers – so they probably won't count.

You must have seen the news of Richthofen's injuries in the papers – or is it not published now? He picked up a violent ricochet in the back of his head about ten days ago (I think it was on the 7th) – he just barely managed to land and mutter some curses, then he lost consciousness.[113] Now he's sitting in the hospital in Courtrai with a huge bandage around his head and he's quite happy again. That is to say, he mostly grumbles about the bad luck that this had to happen to him just now, when he really wanted to get started with the new squadron. Of course, Kohlschein's portrait can't happen now either.

I'm about to get moved again: *Staffel* 29 is supposed to be moved from the 6th Army up to Flanders to the 4th Army. We are expected to arrive in Handzaeme, located between Torhout and Dixmuiden. I'll write to you from there, even if the letters can only be short.
Faithfully yours,
Erwin Böhme

Jagdstaffel 29, August 7, 1917
Dear *Fräulein* Annamarie!
Finally my first greetings from Flanders! For me, Flanders rhymes with "wandering,"[114] and there's not time to rest and write a letter. We've had to move around twice already.

On July 30th, they used huge ship cannons to shoot us out of our original airfield. In the afternoon, I flew to visit Richthofen for coffee – he's now back with his squadron, but he can't and shouldn't think about flying himself for a long time. When I came out of the thick rain clouds on the flight home in the evening to land, I saw right away that a munitions train blew up near our airfield and the train station started to burn, then a bunch of friendly bomb craters also emerged between our tents. That was a nice surprise! But then we were able to fly away with all of our machines in one piece, despite continued shelling and the heaviest rain, and bit by bit we recovered all the remaining material.

First we stayed at Ghistelles for a few days and now we are living at Torhout, but here we have to deal with a very poor quality, wet airfield.

The English seem to have big plans here on the coast. They come in huge swarms. Air battles with mass numbers of planes take place almost every evening – mostly without any significant results, since the English with their new, very fast types of machines are difficult for us to reach. In the long run, that's quite unpleasant. We are always "at a disadvantage" because of the lower speed of our machines. We want to make up for the fact that our opponents outnumber us – but we can only do that if our machines are at least as good as theirs in terms of quality. We should be getting new ones soon.

Flying over Flanders offers many beautiful sights. From Torhout you can see the sea from a low altitude, and when the air is clear you can also see the English coast and the ships between Dunkirk and the Thames estuary. I also see the nice old Flemish cities like Bruges, Ghent, etc. more often now.

I was very interested in what you told me about your work and your professional worries. One must confront winter with some apprehension. There will be a lot of things missing that until now one had been able to find only in forgotten blind spots. Poverty may be particularly bad among the poor in the big city!

Blessed is he who lives on the land! Nature is usually kinder than the war offices with all their regulations. At least my mother has a good stock of firewood at the ready, because only about a quarter pound of coal will be available in the coming winter. Thinking about this, Africa becomes more attractive again.

In the enclosed picture you see a French parasol aircraft[115]– but the French markings have already been replaced by German ones. It recently landed quite harmlessly on our airfield under a cloudy sky. The two gentlemen had mistaken our airfield for Dunkirk and were extremely surprised when we gave them a warm welcome. We all laughed until tears streamed down our faces at the ridiculous situation. They even brought us a very old bottle of cognac!
Best wishes and loyal greetings,
Erwin Böhme

Jagdstaffel 29, August 17, 1917
Now don't be angry with me, dear *Fräulein* Annamarie, for dictating my letter through Erika this time. You're not shocked about this, are you? My relationship with Erika is really quite innocent. Germans don't usually take their "secretaries" with them to the front – but my typewriter brand is called "Erika."

I have to use my Erika today, because otherwise I would have to write left-handed, and you probably wouldn't be able to decipher that. But I'd have to write left-handed because I caught a bullet with my right hand again.

It was just a week ago this evening when, while I was shooting at an enemy artillery plane, some stupid single-seater snuck up on me from below and grazed me across the back of my right hand. The wound isn't so bad – at most it will leave a small blemish. The only unpleasant thing is that the index finger's extensor tendon is slightly lacerated.

This is fatal for someone who wants to lead a squadron, because if I were able to fly again if necessary, I can't shoot with my thickly wrapped hand, and there's even less of a chance that I could clear gun jams. Well, running the squadron is still possible, even if my signatures with my left hand look a bit strange. Incidentally, this is also proof of how "one sided" our training is, that if something happens to a person's right fin, they can no longer even write their name correctly.

My mishap would actually be a very good excuse to move home again. But I can't leave the squadron now. We're only at half strength at the moment. Unfortunately, my top three guys have already been shot down. The day before yesterday, we buried the one on whom I placed the greatest hope.

You should see the Ypres area from above! There is probably no other comparable crater field in all of geography. It's a region of horror. If I were a Berlin Secessionist,[116] I would know where to find my milieu.

Yesterday Richthofen secretly shot down another Englishman. You've recently received a report about how his squadron stood at 199 kills. Well, Richthofen took off quickly to make it 200 – otherwise, he had nothing against the Nieuport. But now he's afraid of the doctor, because he's not allowed to fly again, and the high authorities want to send him on leave again. That's good too – his life is too important and valuable for the German cause.

To my delight, the mail has just brought me your poetic greeting from the Zwischenahner Lake, where I once spent a happy day with Gerhard from Oldenburg a year ago. Are you yourself the poet of these nice verses? If so, I have to ask for the favor of making your "early works" accessible to me from time to time.

Hopefully my Erika didn't write too unorthographically – one is also used to typing only with the right hand.
Many warm greetings with the left,
Yours, Erwin Böhme

August 18
Today I just got some news about the letter I wrote yesterday evening.

Just now, quite surprisingly for me, the message came from the Air General that I have been appointed commander of "*Jagdstaffel* Bölcke," and that I have to start marching over there immediately. So I'm going to "get marching" this afternoon, that is, I'll fly to my new squadron. It's not far from here, in the Ostend hinterland, near Ghistelles.

The new task is an honorable one, but it will also be quite difficult for the time being and will require a lot of work. Of course I'm happy to return to my old squadron, which I already liked because of its name, even if I can't find a single one of my old buddies there anymore. And it is a sign of great trust that the squadron has been handed over to me in order to bring it back up to peak form. This is because its former glory has faded completely in recent times, and there's actually nothing left of the old squadron, except its glorious name. Now it has to be filled with the old Bölcke spirit again. God give me the ability to do this!
E.B.

From a sketch of the history of "*Jagdstaffel Bölcke*"[117]
By its last leader, *Rittmeister* Carl Bolle:
"On August 1, 1917, the squadron began its migration north and, with a short stop at the 6th Army, it arrived in Flanders with the 4th Army. On August 18, *Leutnant* Erwin Böhme, Bölcke's old friend and comrade-in-arms, took over the leadership.

Two prerequisites for new successes were now given: a rich field of activity and a distinct personality as a leader. Now it was time to create the third thing: a self-contained combat group. Here the work of the leader had to begin. Ever since Bölcke's and Kirmaier's deaths, the squadron had always been dominated by one person or a few who towered above their comrades and gave their activities and successes their personal touch. The unity of the squadron and the performance of the others was put on the back-burner.

Böhme was called upon to correct this unhealthy development and to continue Bölcke's work in his spirit. The squadron's ever-increasing performance up to the end of the war owes much to his work…In Böhme they had the leader who was able to make them worthy of their old name again."

Jagdstaffel Bölcke, August 29, 1917
Dear *Fräulein* Annamarie!
Before the start of your vacation time, I would like you to get a warm greeting from me – of course it can only be brief. This is because, first of all, we are doing an airfield relocation today (in the region of Varsenaere, near Bruges; they kicked us out of Ghistelles again), and I'm therefore a busy human being. And, secondly, they operated on my wounded hand again yesterday, where they discovered a very fruitful lead mine.

That's very annoying, because now I have to give up flying, at least for the time being, and this is something that I was recently doing happily again – of course, we can't even mention the possibility of being able to fight.

In my squadron I have decent people, and so it's going to be a sensible operation again. Unfortunately I lost a couple of the most capable men a week ago. I'll see about replacements, hoping to recruit good people for the squadron from people I know from Valenciennes.

The Böhmes can now play a three brothers' game of skat in Flanders.[118] Brother Martin is in Ghent with his giant crates, Gerhard has recently been very close to me again; only our eldest is now in Berlin for a translator course. Rather than skat I'd rather play the violin with Martin, but now unfortunately I have to do without it due to my bandaged hand. I'm indeed able to play the violin with some skill and musical understanding, but Martin is an artist.

We've flown a lot over the sea now. First of all, this is glorious beyond measure. Second, you can sometimes sneak your way over to the other side from behind.[119] And thirdly, one need not pay attention to stupid flak while at sea. It doesn't really do much when it hits, but it always disturbs a pleasant trip. I hate this institution [of flak] – not because I'm afraid of it, but because this disgusting earth worm just doesn't fit with the lofty world of aviation at all.[120]

I recently took brother Gerhard on a reconnaissance flight in a good Rumpler two-seater to show him the sea from above. I flew far out to sea over Blankenberghe against a strong northwest wind. First I went under a thick cloud cover, then I pushed upwards through it, and now we floated in the pure sunshine over the sea of clouds. The real sea could only be seen from time to time through holes in the clouds – from this great height the heavy swell appeared to be light ripples on the immense surface of the water. My "little" brother was totally impressed and delighted.

I then flew south in a wide circle towards that town of Nieuport. Just before we reached the mainland, I suddenly saw – and we were flying against the sun, which was very blinding – a suspicious flashing above us and soon I realized that there were three English Sopwiths, these extraordinarily fast and maneuverable little fighters. They spread out in order to aim at us right in the middle of them.

This was a critical situation. I felt hot and cold at the thought of being responsible for my beloved brother, the father of a family, and in this case, someone who just wandered into this battle. My heavy Rumpler was no match for the Sopwiths. I couldn't shoot or fight at all with my hand bandage – so that meant: get out of there as soon as possible! In a steep glide I dove into the cloud cover, and thus we were lucky to escape our pursuers.

I was so happy when I saw Gerhard at Roulers, where I dropped him off with his troops and, against a sandbag of equal weight, saw him standing safely on his rider's legs again! Never before have I become so aware of how heavily the pilot in a two-seater is burdened by his responsibility for another, while in a single-seater he only has to take care of himself, free and unencumbered.

I'm being called up again. So: happy holidays! Greetings!
Yours, E.B.

Jagdstaffel Bölcke, September 10, 1917
Dear *Fräulein* Annamarie!
Don't be angry that I haven't yet sent home a vacation greeting to you. But my hand still forces me to use the typewriter, and it may be a good substitute for what the head has to say, but not for what comes from the heart.

Since I'm only half-fit for war due to this stupid hand, I want to use my involuntary idle time to bring some squadron requests directly to the source in Berlin. On this occasion, I hope I can get to the Hubertus Mill and finally see you again after more than a year.

From my heart

Yours, Erwin Böhme

Jagdstaffel Bölcke, September 21, 1917
How happy and thankful I am thinking of sunny days with you, *Fräulein* Annamarie! But today I just want to give a brief report.

My return trip to the front went fairly smoothly. I had a fat war profiteer for a sleeping car mate. This disgusting guy annoyed me. Because I could sleep properly, I snored a bit now and again – which I don't think I've ever done before in my life – until the guy started to complain. I had a hell of a lot of fun with it, and I was rewarded for it, as this revolting guy was nice enough to leave the sleeping car a few hours before Düsseldorf. Was that malicious of me or was it good that I punished the bloodsucker? In any case, I was able to ponder my thoughts much more calmly afterwards.

In Düsseldorf I only had a chance to get breakfast and I was greeted by Hauptmann Zander, with whom I made the flight to Valenciennes and had my sister-in-law escorted to the airfield with her small people.[121] I took Zander's mechanic with me as necessary ballast.

From Cologne, we went against a strong westerly wind, so it took me more than three hours to get to Valenciennes, while we only flew one and a half hours to get there. From Düren we used our compass and watch while always flying above a closed cloud cover. I only got out of the clouds again ten kilometers before Valenciennes. Suddenly, my petrol ran out. However, I was just about at the airfield and went into a non-regulation flat glide. Isn't that stupid? From Valenciennes I went straight to Flanders in a single-seater.

Since this had gone so smoothly, fate enviously took revenge with some bad news. In Valenciennes I had already learned that Wolff, who was so skilled and difficult to replace, had died,[122] and now the sad news awaited me that my squadron had just lost *Leutnant* von Chelius, a brave comrade (from the Hussar guards) who had only recently joined us.

On my first flight the next morning (the 19th), I immediately avenged his death by shooting down an English two-seater near Ypres. Another one, number fifteen, was sent down today. And both times I had a piece of your Swedish chocolate beforehand – so actually you now deserve a war medal! My comrades envied me more than a little bit because of this miracle medicine – but they didn't get any this time.

Be grateful that you're with everyone at the Hubertus Mill, heartfelt greetings from your
Erwin Böhme

Rewahl in Pomerania [*Pommern*], September 23, 1917
Dear *Herr* Böhme!
Today is Sunday again – but how differently it looks

compared to eight days ago! At that time we were expecting you at this hour in the Hubertus Mill – today I'm sitting on a Pomeranian beach and, still on duty, I'm looking through a small window above my desk at a large dormitory in which 25 boys are supposed to be having a two-hour midday rest. Only the red rose that you gave me on Monday morning is still on my desk and reminds me of that time.

I guess that I have to explain the whole situation to you first. Here in the small Baltic Sea resort of Rewahl is the "Crown Princess Cecilie—Sea Residence," a children's convalescent home that is looked after by the sisters of the Paul-Gerhardt-Trust of Berlin. Now, in the off-season, around a hundred students from the Strausberg welfare institution in Berlin are housed here – not exactly the flower of German youth! One of the sisters, an old friend of mine, asked me to stand in for her for a week. I didn't want to refuse her, so I drove here on the 19th, was in the house in less than five minutes and already under the bonnet.

Can I give you an image of these Strausberg kids? These difficult little individuals, some of whom have already been convicted and who are otherwise subject to strict institutional discipline, live here in freedom and enjoy this happiness from which they've long been deprived all the more intensely. But they are only too happy to step over the line, partly out of cockiness, partly because of malicious pranks, and partly because they can't do anything else. On the way down here, someone in Szczecin pulled the emergency brake while the car was maneuvering! When a boy on the beach gives me a piece of flintstone that he's found, before I've even closed my hand around it another cute little guy walking next to me has already stole it away and put it in his pocket while I'm not looking!

The supervision of such a group is a difficult matter. We can hardly issue real punishments, certainly not the beatings that are given out in Strausberg. The stick I walk around with up and down their rows of chairs is more of a sign of my status than for discipline.

Things are still good here because sister Lisel, who I represent, did a great job with her group (we each have 25 students). She's pulled out all the good and soft sides in these partly sad, partly stubborn, depraved boys' minds, so that despite all the problems, one often finds a bright, joyous side of these boys. And in any case, my knowledge of human nature is greatly enriched here.

If the boys leave me in peace for long enough, I would like to describe a day here in the residence. At 8am I step into my hallway and as soon as I say "good morning" everyone jumps up like one of those roly-poly toys [*Stehaufmännchen*]. Getting dressed is no problem. Then everyone has to make their bed, and if it's not perfectly smooth, they have to do it again. At half past eight it's time for breakfast. Three cups (the boys call them "bowls") of soup for everyone, two hearty slices of bread, and today, for Sunday, two rolls, crunchier than you've seen them in years, and then we let the boys go sliding, or what they call "Strausberging."

From breakfast to noon we play on the beach or in the forest, depending on the weather. So that means: be careful! You always have to count the group again, because someone always likes to "sneak out." "I'll rent out the farmer, then I don't have to go back to Strausberg" is a popular theme.

At the table there are mountains of the most beautiful food, like it's not at all wartime, and then comes the two-hour midday rest, during which I can breathe a sigh of relief. At 3pm we get coffee with milk. In the afternoon we usually go for walks in the surrounding villages. "Look, the beautiful apple trees!" says one, and "Attack!" shout the others, and we have trouble "holding" them. But we also get them fruit as a gift.

Only when the boys are in bed do we get some quiet and sneak down to the beach again to enjoy the roaring waves or the quiet breathing of the sea, which we hear nothing of the day before because of the noise made by the boys.

The day before yesterday we saw the Baltic fleet, eleven large ships, on the high seas, but clearly visible with the eye, lying at anchor here for a few hours. The planes fly from the coast several times a day at very low altitude. Did you also fly over the sea again?

I still wanted to write so much to you – but things are getting lively over there in the hall.

Sunday and Monday were too nice!
Sincerely yours,
Annamarie B…

Jagdstaffel Bölcke, October 5, 1917
Dear *Fraulein* Annamarie!
It's terrible that I haven't thanked you for your cheerful letter from Rewahl. But in the meantime we've played gypsies again and moved from Varsenaere to the Roulers area – but this time we weren't just forced out of the airfield. We swapped places with another fighter squadron.

In addition, there was a second thing that blocked me from writing a letter: I've lost your new address. I only know that your cheerful convivium[123] from Sonnenau has moved, and that it also ends in "au" – do all of the streets in Hamburg actually rhyme with "au"? But this matter isn't amusing to me at all. It's actually a great worry for me that I don't even know whether you'll be able to find my thoughts. I hope this greeting reaches you at the previous address.

So please, don't be angry with me, otherwise I'll be sad,

and I don't want to let the beautiful, wonderful, memory of my visit to the Hubertus Mill drift away. Otherwise I'll have no choice but to go to Hamburg in the near future to refresh this memory.

I would have liked to have seen you in Rewahl amongst this pack of badly-behaved boys!

But tell me: do you count something like this as one of your vacation breaks? That's *Fräulein* Annamarie for you again – even sacrificing your free time for others!

For the last five days we've been in Rumbeke, near Roulers, not very far from the Ypres salient. The artillery barrage is terrific again today. They also shoot as far back as Roulers – but our place has only gotten a few random hits so far. We live here in very nicely furnished barracks. I even have a whole little house to myself.

Unfortunately, my squadron has shrunk to about half in the last few weeks due to injuries. One was shot down, and a lot of housekeeping was done on my part. I miss *Leutnant* Wintrath, who was shot down over the sea, and Vortmann, who was badly shot in the hip. Unfortunately, Plange is still on vacation, peeling potatoes at home. I've recommended to four other gentleman that they should swap fighter aviation for another occupation. The rest will be fine. I'm busy recruiting again.

Voss should not have fallen![124] This is a heavy loss for aviation. He was my comrade-in-arms in the old Bölcke squadron and was now Richthofen's right-hand man.

I forced down number sixteen this morning on an empty stomach. This time I was successful without your miracle chocolate. It's actually long gone. I think your *Fräulein* von Wiese, with her many connections, could go to Sweden again soon and bring back something nice and wonderful from there.

Hoping faithfully for a gracious letter soon from his esteemed girlfriend,
Erwin Böhme

Hamburg, Wartenau 10, October 17, 1917
Ten enemy planes and a captive balloon were shot down yesterday. *Leutnant* von Bülow brought down his 23rd. *Leutnant* Böhme forced his 20th to crash after an air battle.

The evening newspaper just brought us this surprise with today's army report, which spread like wildfire among the people of Wartenau. We are now celebrating your great success with lute sounds and clinks of glasses filled with Overbeck Rhine wine, and we wish you continued success and victory!

Minni C., Elisabeth H., Lotte von J., Gertrud O., Anny von R, Theo P, Annamarie B.

"With the Aces,"[125] October 20, 1917
Dear *Fräulein* Annamarie!
So, things aren't so bad after all!

Today, after a long delay, your long-awaited letter from last Sunday finally arrived. I read it in the air, because we were just getting ready to take-off when the mail arrived. Now I finally know where to find you! So I want to send my greetings to you right away and make you jealous – because now I'm sitting in a cozy, heated room, while they don't want to give you any coal before Saturday? Barbarians!

At the moment I'm living like a prince again. Although it's called a barracks and looks directly at the airfield, it's in fact a very smart little house, à la Hansel and Gretel, with a front garden, a glass veranda with a view of the artillery barrages in the Ypres salient, living room and study, bedroom, changing room, and bathroom. At the other end of the airfield are the residential barracks of the other gentlemen, each with eight single rooms and the mess hall – all very nicely and amicably furnished. It's just a pity that we'll probably have to move on again soon, since the envious Englishmen have already flown over our airfield to take shots at the neighboring town. But we'll take everything with us – except the garden.

Just think: I made the acquaintance of a merchant, this time an Englishman, as I did the other day in the sleeping car. His name is Mr. Ortweiler, some of his family comes from Frankfurt, and he is currently full of hate for Germans. I met him on the front the other day (on the 16th) in an English Nieuport single-seater in which the friendly gentleman was sitting. Of course, I invited him to visit us. When he didn't quite want to right away, I became a little bit more aggressive and forced him downwards. He kept waving his hands at me, insisting that I didn't really want to hurt him. He ended up on our airfield completely healthy. We laughed uproariously. This was just the kind of piece that I was missing in my collection. It was also number twenty. I got numbers seventeen through nineteen on the 10th, 13th and 14th.

I've really enjoyed my squadron now that I've unloaded the dud pilots. They just fly everywhere wildly and start shooting. This afternoon there were two big battles between single-seaters. With the second, over Passchendaele, I had a damn gun jam again, and it was imperative for me to shoot – I lost another capable comrade flying next to me, *Leutnant* Lange, whom I could not get out of trouble. But the others shot down another three Englishmen.

My brother Martin has now been over England a few times – they've really started to heat things up over there. The day before yesterday, he was with me. We made beautiful music. The drumbeats of the bombs became quite fitting as accompaniment to the violin.

Unfortunately, someone recently broke my car that I use for travel. After I reported the loss, I also learned on top of everything that I actually owned it myself, and figured out how that came about, etc. I pressed three different rubber stamps on this letter – now I'm feeling better.

Your faithful friend wishes you the same,
Erwin Böhme

Jagdstaffel Bölcke, October 24, 1917
Dear *Fräulein* Annamarie!
His Excellency von Hoeppner[126] has commissioned me to deliver the greetings of the flying service to the parents of Bölcke on the anniversary of his death.

I hope that I can make it from Dessau[127] to Hamburg, even if for a very short time.

Always yours,
E.B.

Telegram from Magdeburg, Oct. 28, 1917
Arriving tomorrow, Monday afternoon, at 1:26. Please wait at Wartenau.
Erwin

Rumbeke, October 31, 1917
Dearest brother Gerhard!
I now have a fiancée! You, my dear brother, should hear this story right away. I got back from the Dessau trip today and from there I was in Hamburg for half a day. I arrived after 1pm, and at midnight my train went back to the front. In these short hours I received my happiness from heaven. It still has to remain a secret until Annamarie's parents have given their approval.

Of course it's Annamarie from the Hubertus Mill that I've often told you about. Ever since we did that famous "emergency landing" back in May, we met each year and engaged in correspondence. For me, it was love at first sight and, as she confessed to me, it was also for her. But nobody dared let down their exterior, and we were each careful not to betray our hearts in our letters – she out of girlish shyness, and I in my own reserved way. We were both tormented by the uncertainty of how things looked in the other one's heart. Now you understand the heavy mood that dominated me this summer. Now I had to finally clarify these things.

Annamarie is exactly the kind of girl that suits me: a genuinely feminine nature, fresh and natural, without any trace of that kind of modern masculine woman that I hate, and yet she's smart, brave and has the strong will of a man. I also find her beautiful, with her lively dark eyes, from which goodness and loyalty shines, and sometimes a funny joke also flashes out.

Am I just a reckless man? Without a secure job in life, without a definite plan for the immediate future, to chain someone else's destiny to mine! But I firmly rely on the favor of the good spirits who always help the courageous and strong.

Boy, I would have never thought that this old man, who had grown to become serious and strict through life, could be this happy again.

Rejoice, Gerhard, with your old, rejuvenated brother,
Erwin

Letters to My Fiancée

With the aces again, October 31, 1917
To you, my dear Annamarie!
A long, very long kiss first, and then a hasty, very short greeting!

I've just arrived, just in time to catch the postman for a moment as he stomps in.

Strangely enough, the long drive wasn't that long for me – I always saw a pair of intense, loving eyes in front of me, and sometimes, when I'd dozed off a little, a soft cheek was close to mine.

You dear heart, what a rich man I am!

Goodbye until tonight – there are too many people waiting outside who think the war is going on.
Your Erwin

Rumbeke, October 31, 1917, evening
So, my dear heart, now I can come to you during a cozier hour – earlier it was too uncomfortable and restless: vacation seekers, clerks, mechanics, etc., stormed my office. Eventually the English came with bombs, and I went up to strafe one of them right away, but only reached him when he was already heading for the Penates region, near Ypres.

In the squadron I found everything was in good shape, to my great relief. Gallwitz, who stepped in for me, has also meanwhile shot one down, and he's very proud of it; he's still a very young fighter pilot, but absolutely reliable. I couldn't bring myself to hide my happiness from my comrades – there was jubilation! They give all their best wishes and greetings. Besides, you would agree with me that "emergency landings" at home, especially in the month of May, are the most dangerous undertakings.

My brother Martin was very happy for us this morning. By the way, he acted as if it all was simply expected to happen this way – well, he must be right. I arrived in Ghent late last night after a bit of a harrowing drive and stayed in the city for the night. This morning I strolled out to the airfield in thick fog and of course had to kick Martin out of bed first. The "giant" aircraft pilots have already started their hibernation.[128] I wasn't able to fly over to my squadron until about noon.

The first thing that greeted me here was your letter of the 24th with the thoughtful reflections on Bölcke's death. Can you believe that what moves me about this letter is that it's the last of its kind and still addresses me as "you" [*Sie*, formal]?![129]

What a fool I was that I didn't get my happiness to fall from heaven a long time ago, that I didn't ask you the big question weeks ago! That's why I came to the Hubertus Mill in September. But as we were pacing the garden that morning and I was cutting the last rose for you, the fear of getting a "no" constricted my throat so much because I knew that a "no" from you would have so shaken my sense of psychological balance that I would then no longer have been suitable for my responsibility-filled duties at the front. It's strange that this husband, who is not afraid of death or the devil and flies unhesitatingly towards an enemy squadron, is too shy to ask a girl who wants the best for him whether or not she loves him!

My thoughts are always with you, my dear, and this is my prayer: that a way will soon be found so that we can always be together. Please send me some pictures of you soon, new and old, and give them a few kisses too. See you tomorrow, my dearest love.

Your Erwin

November 4, 1917, Sunday morning
Good morning, my dear heart!
Yesterday was a happy day for me: it's the first letter from my Annamarie that says "you"[informal, "*Du*"] But I've been thinking of you with "*Du*" for a long time – actually, from the beginning. But that wasn't quite the right image of you. However, that doesn't matter – in the quiet hours I have your dear face very much alive in front of me; maybe when you think your thoughts go straight to me.

So now you have the difficult mission with your parents behind you. I hope that they approved of everything with an open heart! I'm totally confident that your mother will be helpful with everything. Tomorrow I figure that I'll get a greeting from you at the Hubertus Mill. It's an awfully long time, having to wait for four days to find out what you've experienced. It's a good thing that the war isn't in Africa! We had to wait three week from one postal visit to the next. When the black postman ran 30 kilometers in three hours from Wilhelmsthal and then dumped out his mail bag, it was the focus of all the minds of the white people at that moment, and five minutes later everybody sat in a corner and read.[130]

One more thing, dear! If you really don't receive a greeting from me for a few days, you know that all sorts of coincidences and obstacles can occur with the mail from the front [*Feldpost*]. Yesterday, for example, our motorcyclist, who was taking the mail to the next *Feldpost* office, tipped over and fell while he was halfway there. That was a whole day's delay. And don't be surprised if my letters turn out to be a bit hacked up in style – only seldom can I write one in a single sitting.

Your girlfriends in the Wartenau are real rascals for teasing you about the secretive visit of the *Leutnant*. But their congratulations for me are courteous and kind – so say hello to them for me, and I won't miss another opportunity to introduce myself. The myrtle tree that you donated to me needs your further devotion – I'll justify my next leave with this important necessity.

Yesterday I got a very valuable addition to my squadron: the Bavarian *Leutnant* Max Müller, who with 31 victories is now in second place (after Richthofen) among the living aviators. He was part of our squadron a year ago, but then he spent the summer with another squadron.

Now I have to tell you about my 21st. I had been at the front with my squadron without anything special happening, and during the return flight I broke away to visit Richthofen, whom I had not spoken to for several weeks. After a little coffee hour, I flew back home, but before I landed, I saw that my people were already on the move again, so I flew back to the front, where I soon found them. High above us was a squadron of English single-seater fighters. We tried to climb as fast as we could. Since I had used up most of my gas on the first flights, my machine felt quite light and climbed like Charlemagne.[131] After a short time I was far ahead of my companions and reached the same level as the Englishmen who were still above us, and thus I kept an eye on everyone. Finally someone came up with the stupid idea of attacking me from above. I made their first attack ineffective by quickly flying towards them, so he pulled his plane up and was again immediately about 200 meters above me. He was flying the latest type with a very powerful motor. From then on he made for our five more hesitant attempts to attack, but each time I was immediately under him again, and so he couldn't get a shot. Gradually he lost height, and at a favorable moment I was able to turn the tables – now he's down, the stupid fellow!

The whole affair lasted at least five minutes. Of course, it shouldn't have lasted any longer, because I came home completely out of petrol. Meanwhile, they'd thrown some bombs here, but they only made some holes in the surrounding fields.

There's thick fog this morning, so I'll get going on my paperwork – in the end, everything makes me think about you, Annamarie.

How I look forward to seeing you,
Your Erwin

Hubertus Mill, November 1, 1917
My beloved Erwin!
Now that the sun has gone down, I finally get around to writing you a greeting. To do this, I fled to the red room at the back, where no one looks for me. Mother is sitting in the next room at the grand piano. She's playing the old Luther song as though she's in the mood for a festive Reformation: "A mighty fortress." The second verse is particularly dear to me: "With our power...."[132]

There was a big day of fighting here yesterday. It cost tears and a lot of love. "We" have won – the price is the most precious: the blessing of the parents. Come here, come soon! Bring lots and lots of love and everyone will receive you with open arms.

Mother had expected a sign from you today. I comforted her: You are probably already on your way. At noon a pilot flew over us, but it wasn't our pilot. When are you coming?

Today was a day full of happiness and sun like I've never experienced before. The world looks so different to me now. Everything is so bright. The earth has always been beautiful, and I knew it, but it has never seemed to beautiful to me as it does today in my happiness.

After dinner I went into the forest with my sisters. The sun was shining, nothing but gold was hanging on the trees – we only have to pick it. I think the forest is still echoing with our happy songs. Do you know the second verse of the folk song, "It wanted to sneak in"?

> I heard a little bird whistle,
> It whistles all night
> From evening to morning
> Until the day dawned:
> "You lock my heart in yours,
> Lock one into the other!
> A little flower shall grow from it,
> It's called a forget-me-not!"

Got to hurry – we've got to get to the train. I want to post this greeting in Berlin so that it travels quickly. Sequel to be written tonight on the train. Tomorrow morning I'll be back at work in Hamburg.
A thousand greetings from yours who is tied to you in love,
Annamarie

Flanders, November 5, 1917, evening
My dear heart Annamarie! The postman came unusually late this afternoon, and for this reason, he found someone waiting eagerly for a loved one when he finally brought three letters of greetings: your card of the 30th from the Hamburger Bahnhof, the letter from the red room, and midnight one that poured out of the fountain pen on the express train. So I now have your whole difficult journey coming alive before my eyes.

Really, my dear, did it cost tears, the day of the big fight? Well – success without effort is not happy success, and thanks for already doing the fight for me. After this kind of artillery bombardment preparation, there's no way I can fail. This confidence allows me stay calm as I give in to the necessity of staying here a little while before I can travel to the Hubertus Mill.

So listen, Annamarie: in the next two weeks, duty at the front will not permit me to get free again. First, it's because I'm hiring four new squadron members in the next few days, and then there will be a change of location. And the most compelling reason: there are going to be some special combat operations in the near future, for which we must give our complete dedication.

It's really wonderful who the entire situation up here has recently progressed! You also see this in the army report: "The Tagliamento river has been crossed mid-way, our troops are still advancing" gives us hope for the next few days.[133] Our troops are still magnificent once they can get moving. We have to gain the advantage, not just hold out, and this is also true if we want to make the Erzbergers nail shut their filthy traps.[134]

Meanwhile, I'm busy working on a plan that I want to use to disarm your father's concerns about our immediate future, so that we can then exploit our victory even further and bring my dearest bride home. We'll have to sketch out this plan next time we're together. I think your grove of hazel nuts is particularly good for forming plans – I need an open sky above me to think good, clear thoughts.

Your nightly railway letter gives me great joy. Please tell me often about your little daily experiences – this brings my thoughts very close to you.

You were right to defy the timetable change for the small train and bravely march the long distance from the city to the Hubertus Mill without consideration for the soles of your shoes, which are now so precious – I would have done it exactly the same way. Incidentally, it's not only because of the time gained, but much more from the feeling that nothing improves your inner concentration more than walking alone, especially when it's the last stretch before a destination. Do you feel the same way? Some people think it was funny that

I used to get off a few stations before Holzminden when I went home for the holidays in order to walk in a leisurely way across the beautiful Solling hills towards my home town.

I can also relate to your mood about your improvised church visits. I'm not a church-goer – for that our average Protestant pastors would have to be more full-blooded men than I've seen over the years, especially during the years when a young person is so influenced by external impressions. But one can be religious without a liturgy. I believe that such a church, in which only really religious people quietly come together, could be a real place of worship. Is there such a thing? Incidentally, I have to confess that I became much more God-fearing during the war – it's also possible that it's only now that I've realized that I always was.

On November 6, Noon
Tonight and early this morning there was a "huge battle" in this sector. Despite the bad weather, we've already brought down five Englishmen today. The squadron is developing splendidly. I just bagged my 22nd. I've got to head straight to the brigade headquarters for a meeting.
    In unending love,
    Your Erwin

Rumbeke, November 7, 1917
Dearest! Today it will only be a short greeting. Urgent reports – twice I was called away to fly in the middle of this – don't let me get back to this letter to you.

This morning two Englishman once again lie on the ground on account of our squadron, despite the miserable weather. We had number 22, whom I forced down yesterday, over for coffee in the afternoon – a very nice guy. He also resisted courageously to the very end. He was amazed that we also caught his two escort planes and he only believed it when we told him the names of the occupants. It was a really enjoyable hunt in the clouds, a real game of hide and seek: whoever saw the other first got to shoot.

More and more really chivalrous interactions have now developed between us and the English pilots: we give each other messages, attached to a little sandbag, to tell each other about the fate of the missing, and also to throw down wreaths for the fallen.

In order to eliminate any doubts about lost letters, I would like to suggest that we number our letters consecutively. So this is number 1[135] from my side. We can then always begin our letters like this: "In polite regards to your honored No. 7," etc.
    Heartfelt,[136]
    yours Erwin

Jagdstaffel Bölcke, November 9, 1917
My dear heart!
Flanders is a bad, completely unpredictable country. Today I should have written to you all morning – so much, everything, came to mind when I read your letters last night. But this morning, contrary to what I expected, there's no fog and no rain, and I'm now sitting here with one leg in my plane, and the other in the inkwell, my head in the squadron office where all kinds of urgent things are still waiting, and my heart in Wartenau. Well, you'll see the result of this situation.

So: yesterday evening I was so dead tired that I already had my eyes half-closed while I undressed. But when I was lying in my nest – the sign to the outside world that no one is allowed to disturb me any more – I took your two letters that arrived towards the evening and then talked to you for a long, very long time. Did you feel a sense of that again? I then kept your dear picture under my pillow, and when I woke up this morning you got a very firm kiss. In order to compete with that, the anti-aircraft guns cracked again, and then it was time for me to go straight up into the air. Suddenly, the sun emerging above the clouds brought me your morning greeting.

The new little photo is more true to life than the first one you sent, but it's not quite my beloved girl. Later we'll have to visit my dear Hans Kohlschein, the painting master or master of painters in Düsseldorf – maybe he still has a piece of canvas waiting for us. You must also get to know his wife soon – she's like a sister to me.

You seem to be afraid that now, for your sake, I might be seized with ambition to chase the Pour le Mérite. I've already earned it many times over now that I've conquered my Annamarie. Incidentally, they're not as generous in giving it out as in the days when fighter aviation was in its infancy; Immelmann and Bölcke got it after their eight aerial victories, Richthofen after his sixteenth. By the way, for me all medals only have the value that I ascribe to them based on my own feelings of personal achievement. So it is actually quite immaterial to me whether I have one more or not. But I would be happy about the Pour le Mérite, because it would be an acknowledgement of the brave squadron that I lead.

Your sister Elli is a splendid girl – her letter made me very happy. It's a nice feeling that your other trusted friends also share so warmly in our joy. This sincere mutual joy from people that you treasure is a valuable gift that strengthens our self-confidence. The kind of envy that you find in lesser creatures doesn't affect us. My good comrades also honestly share my happiness and say hello to you every day.

We still want to think about how we can set up the most ideal trip to the Hubertus Mill. My first destination is definitely Hamburg. Will I find someone from the Hamburg War Relief Office when I get off the train? How I look forward to this journey!

Have the little roses arrived from my war garden? I was just able to give them to a traveler for expedited transport. It will probably be the last ones we see this year – after the cold rain, none of them want to emerge any more.

Now the warrior has to get back to his dreadful paperwork. So just one more heartfelt kiss, my love, from yours.

Erwin

Rumbeke, November 10, 1917

My deeply beloved Annamarie! Your dear photo sits next to me, and when the world wants to appear gray to me through the rain-clad windows, I look into your eyes and still have my share of sunshine. As much as I always want to resist it, I am one of those people whose mood and energy are very dependent on the weather and the environment. By the way, I don't even know whether I should see this as a mistake or a disadvantage, because I think that those whose minds are always able to be upbeat in their own way in dull surroundings are unable to welcome the arrival of spring. Well, once you're with me, it will never be so gray again.

Tell me, dear, is your father very worried about our plans for the future? I have to admit to my own surprise: I'm not so worried about it! You always have to give yourself a little bit of time, especially nowadays, if you don't want to make impractical plans and then experience disappointment. If the "farmhouse" doesn't work, then we'll find something else – it just has to be an area that resembles Berlin as little as possible. I'm putting out feelers to several contacts and will tell you more about it soon.

Right now I'm the deputy (not a permanent position!) commander of the Ypres fighter group [*Jagdgruppe*]. That includes the three fighter squadrons brought together in this sector, since the regular commander (the senior of us three squadron leaders) is on leave. As such, I have to regulate the deployment of the fighter squadrons, but in addition I have a whole house full of daily reports to finish – a special treat for me! In addition, I'm getting used to reading in bed, since important phone calls often come in the first hours of the night.

At this moment, I'm reading the letters of Goethe's mother with great pleasure.[137] Do you know her work? That was a splendid woman! You can learn much practical wisdom from her, and it all comes out so comical and natural, never like a lecture, and that's why it has an immediate effect. It's a fact that in the development of important people, the most important thing is usually the origins of the mother – but a happy mixture is of course also part of it.

On this occasion, a doctrine of faith: the attempts by modern women's circles to appropriate all domains previously held by men and make it their sphere of activity, and to eliminate all differences in occupations, etc., is not only extremely repugnant to me, but I also consider it a great danger. If only because the man, for his part, cannot conquer the territory of the woman, and thus the corresponding balance is missing. But above all: the difference between the sexes is a God-given one; they should complement each other, but not try to be the same. You know it, Gorch Fock[138] knows it, and every human child who is of sound mind knows it: "A whole person needs two." And that's why you don't need to be afraid for the future – nature can't be bent over the long-term, and anyone who tries to do it has to become an unharmonious and unhappy person. We all admire and are proud of what German women are doing now in time of war, offering their adaptability and their strength. But what is required by necessity should not become the law during normal times.

We two, dearest Annamarie, always want to sing along with Walther von der Vogelweide[139]: "I am yours, you are mine" – everything else may come as it must.

Our motorcyclist is about to drive or slip away with the mail. Will he also bring something to me from my girl? I think so!

Let yourself be kissed, my love, by you – your Erwin

Rumbeke, November 11, 1917

Dearest! From a half-cleared out abbey, a quick greeting and thanks for 4 (I'll write it out: four) dear messages, which the friendly postman brought to me from you all at once today. We hurry to prepare for the move, and tomorrow we'll escape from the intruding cannons – our place is being fired at from over there with some large caliber artillery. As punishment, we chased a whole squadron of single-seaters to Ypres, and Max Müller shot down one of them. But it's not exactly comfortable in our dwelling at the moment.

So it's true that we often talk in the quiet (except for the guns) hours of the night! For my part, I wanted to ask you yesterday whether you need some shoes. Will now try to get some soft ones in Ghent. One must now look for sources here too – I hope that Brother Martin knows something. What if they don't fit afterwards?!

Now I have a travel plan for the near future: you'll meet me as far as Düsseldorf, where we could both arrive in the morning. We'll stay there for a day with Ella Kohlschein, and the next morning we'll drive to my mother's, where we'll arrive at noon. Then we'll get back on the train at night, arrive in Berlin in the morning, and from there we'll soon be with all of yours at the Hubertus Mill. Do you have any concerns? If so, we can do it differently. But meanwhile I'm going to crawl exhausted into my nest with this plan and paint it in bright colors. Good night, my heart – one kiss and then another!

    Your Erwin

In a cold room on the new airfield,
On November 14, 1917, early
Dearest Annamarie!
How fortunate that the temperature of the heart does not depend on that of the hands! Otherwise, you would get a cold greeting today – so it will be short. Flanders fog is really disgusting if you can't heat it up. Thanks for Numbers 11 and 12! I'll answer in more detail as soon as my furnace starts working – this evening, I think.

So yesterday I was in Ghent to buy shoes. Not very good prospects! Nothing but appalling garbage at even more appalling prices. But Martin knows another source where you can order them, made to order – so please, send your measurements, preferably from the cobbler.

Butter?! Our buyer is on the road at the moment and will probably be back from his sneaky, black market trade trip in a few days.[140] Then I'll send it off immediately, but I'll try it out on a small scale first to see if something like that will get through – most of it is now being stolen. Occasionally, I'll give guys on leave something to take with them. That's the safest form of transportation right now.

Your mother wrote me a very kind letter, which made me really happy. I've just asked her whether she has anything against the travel plan to meet in Düsseldorf – but I don't even have your approval yet. Will she come? Burr, this pen!

    A hot kiss with cold lips!
    Your Erwin

November 14, 1917, evening
My dear sunny girl!
When this greeting reaches you, did you already receive my frosty letter from this morning? Now my furnace is working and I can live in this room. For the time being, we live quite far away from the airfield in one of the larger farm houses. The room is small but clean, and it has such wonderful, beautiful murals of rural art – I like them sometimes better than our Secessionists in their bland coffeehouse brilliance.[141]

If you meet me in Düsseldorf, then let Ella Kohlschein show us some good stuff. Hans Kohlschein himself is currently a war artist in Warsaw and paints Poles,[142] and most of his more recent paintings, especially his fabulous war compositions, are currently on display in Berlin. But there are still all kinds of things in the house from earlier years that you will like. Get the last October issue of the Velhagen monthly magazine.[143] You'll find a long essay about Hans Kohlschein with is paintings.

You recently asked if I enjoy beautiful shapes and vibrant colors? It's far too much for me to think that the general taste of our own bourgeoisie, which is so focused on utility, could amount to something! As soon he gets the idea to bring in some color, it becomes tinsel and kitschy – not only in painting and related things, but also in our music today (see the tastes in Berlin operetta!). Well, we're definitely not moving to Berlin, my farmer's wife – right? There is good music elsewhere, and by the way, we can make our own.... Oh, this pen! Can't you send me a few soft, not pointy pens? But it's not too important – I'll be there soon, and then we'll go shopping together.

Thanks for the kisses wrapped in May flowers! Yes, I'm also thinking very much about a certain day in May. That's so long ago now. But it seems to me as though we've been very close ever since that day...

Early on the 15th
The lights suddenly went out last night, so I couldn't write any more, and this morning I have to scope out airfields. So for now just a heartfelt, good morning kiss from your Erwin.

Bavikhove, November 16, 1917, evening
My dearest fiancée!
I complained too soon just now – your dear Number 13 arrived, a day late.

Do you know why this letter in particular makes me so happy? Because it shows me that my girl is brave before God. Just yesterday, someone who certainly meant well wrote to me that I would have to stop flying at the front after all – that I owed it to my fiancée. But you've already given me the answer to this in your letter today. Thank you, my dear, dear heart. We both know that we cannot master fate after all, and we also probably feel better, humble creatures that we are, in doing what we have to do now. I know that you are always with me – I don't think about anything else. I feel sorry for anyone who finds peace of mind in just taking care of their house during this time, because they will never be able to know the greatest feeling of happiness afterwards. I know that we both understand each other completely on this.

Dearest, I long to see you soon. The day is getting closer now. I can't say exactly when today. Our operations are getting somewhat complicated right now.

Yesterday morning, when I was looking for an alternative airfield, a lonely little cornflower waved to me. It really wanted to go to you. Now I'm sending it – it will bring you a heartfelt kiss.

But I also found something else on this trip: in a very small village, in a very small shop, a very small pair of shoes were very lonely – I bought them at the risk that they might not fit you. A friend who is often proven to have life wisdom told me: nowadays you have to choose your fiancée according to the shoe size that you have at the moment – not the other way around. In spite of that, I dared to buy them anyway. However, I do not dare to send the shoes in a parcel, because while it's underway, someone might get interested in applying the above-mentioned formula. So be patient and don't set your expectations too high.

By the way, you also wanted to know when the writer of this letter was born. According to the official document, on July 20, 1879, my parents mentioned in the Holzminder newspaper's "daily announcements" that to their joy I saw the light of day as their second son. Supposedly I also cried out horribly, but I never believed that – it's just not my style.

Early on the 17th
Think Flemish fog this morning.[144] You can barely see ten paces away. One of our trucks confidently drove straight ahead at a bend in the road and is now lying helpless in the ditch. We are very happy about the weather – at least now we can check our machines thoroughly.

Personally, I would of course prefer to be in the Swiss mountains. One of the most wonderful things I have experienced is flying out of the lake basin, which at this time of year lies in a thick fog for weeks on end, up to the sun and bright mountain peaks. The mountains then all emerge from an almost flat sea of fog and the sun glows all around them – you can't even imagine that people are swarming around down there in the gray haze talking about the ugly season.

You and I always want to look for the sun!
Your Erwin

Bavikhove, November 19, 1917
My dear heart!
Did I really boast so horribly about my heavy workload? That was not my intention and it doesn't reflect reality. It's only during the absence of the fighter group commander on leave that I have to accept the problems otherwise intended for him and administer the deployment of three fighter squadrons. Plus there's all the paperwork, sometimes more, sometimes less, and I sometimes lack the will and patience for it all. But otherwise it's doable.

You ask whether or not I have been doing any work with Richthofen. No, not for a long time! He is commander of a fighter wing [*Jagdgeschwader*], which consists of four fighter squadrons [*Jagdstaffeln*], which also belong to our army and is quite close by, but are always available to the army command and are for special occasions, while we have our group assigned to specific sectors of the front. Of course we don't have to be too concerned about this – we often hunt in the neighboring sectors if the situation arises. That's the thing about fighter aviation, that you find new tasks with every flight and then do them according to whatever decision you need to pursue at the moment.

Yesterday afternoon on my flight home from the front I landed to see Richthofen for a coffee– they always have such delicious cakes there. Richthofen is now constantly being stalked by painters who want to do a portrait of him. Yesterday he said that he wanted to give up flying completely and only focus on having himself painted, since that was less dangerous and yet would make him famous at least as quickly.

I'm not really in favor of publishing our "actual engagement" in the newspapers. I'm thinking: this time we can put an avalanche in the mail and send announcements to all our friends. We can simply set aside a whole morning to write addresses. One of us writes, while the other supervises, and then we can switch. With every switch, the one who gets relieved gets a kiss. Agreed? That's suitable for war.

Today, I think, I should get your answer to my travel proposal. I'm really looking forward to it, and I really hope that you are in agreement. I don't yet know how many days I'll be able to spare. People are pretty stingy here, and besides, I want to keep my chance at getting a real holiday around New Year's. Everything will work out smoothly.
With yearning love,
Your Erwin

November 20, 1917
Dearest! I've just come back from my flight and see that the postman who was dispatched earlier hasn't actually left yet – now he can carry one more greeting. I've since earned another bite of your chocolate by snagging number 23. He was about to attack one of our reconnaissance planes – now he's fallen on the floodplain West of Dixmuiden.

Your Number 19, in which you take a close look at my Düsseldorf plan, came last night. Oh dearest, I'm afraid you'll experience some surprises with me in regards to observing the traditional ceremonies – I want to make an effort, but it's so difficult! It was only afterwards that I realized that traveling, especially at night in the unheated trains, is not pleasant at all, and that it's very stressful for you – of course a guy who wants get together with his girl doesn't think of such things. But can I pick you up in Hamburg?

This morning I was finally able to dispatch my courier with the care packages (in the true sense of the word) to you.[145] The messenger is visiting his *Leutnant*, a member of my squadron, who is in hospital in Altona – this way you can be sure to get these precious goods acquired in war. Everything tastes good – with the exception of the shoes! I'm sorry that the butter jar isn't even full, but I couldn't find it anymore. Incidentally, the ancient Germans didn't always have butter either, and yet they thrived. But once we have our own cows, then my dear farmer's wife will swim in butter to her heart's content.

With true love, your Erwin

Bavikhove, November 21, 1917
My beloved girl! Thank you very much for your picture in the nice frame! You've helped prevent a real emergency by giving me this, because I had not yet succeeded in finding a suitable frame. Now you're standing in front of me on my desk between a few little Flemish flowers. By the way, it's a great advantage that there's now glass over the picture – it doesn't suffer that much if a kiss turns out to be a little too strong. Do I have to send you back the little pictures from your golden childhood days, or can I keep them?

Enclosed is a quick return gift. Yesterday when I was rummaging around I found a whole stack of photo postcards from "our successful fighter pilot *Leutnant* Böhme," based on a photo taken by Ella Kohlschein; they are made by a Berlin art dealer who has published a whole series of "Heroes" postcards.[146] I'll send you a dozen right away, so you don't have to strip the decorations from the Hamburg shop windows.

Today there was also a good letter from your dear mother. She, of course, isn't thrilled with my nice travel plan, for reasons which, for good or bad, I have to acknowledge as fully valid. I always seem to forget to respect social conventions! Maybe I'll learn that too. Since I thus have to reject my plan to pick you up in Hamburg, we'll only see each other again at the Hubertus Mill. But maybe you'll come to meet me in your thoughts, dearest?

Among other things, your mother writes that my father has strong hopes for our future, considering my prospects in Africa. I'll have to go through this whole thing about our African company with him very soon. A lot has changed over there because of the war. Above all, I don't intend to spend the best years of my life there, and I'm even less inclined to transplant my German girl below the equator.

I don't regret my six years in Africa. They gave me a lot through the happiness of being able to create freely and the close connection to untarnished nature, which I love as a nature lover; they also significantly expanded my entire field of vision and my knowledge of human nature. What a source of pride and joy I felt when I saw the long suspension railway bridge, which I built, from the Usambara line up to the Pare mountains – working reliably! What a powerful experience it was when I went in the dark into the Schume forest to the lonely place where, in the evenings and the early morning hours the strong buffalo like to hold council!

That's all well and good for a few years. But in the long run a German can only thrive where there is a change of summer and winter gives you fresh strength. Believe me; in those years I missed nothing more than the times when the reinvigorating god of winter sent his first scout, the hoarfrost, and I missed the snow crunching under my feet. A permanent stay in Africa is draining. For me, it's the result of long observation that I immediately associate the concept of the "old Afrikaner"[147] with the idea of something that is inadequate. Even the natural wonders there are overly animated for me – everything is like pepper and vanilla. It's like wanting to be constantly fed in a cake shop [*Konditorei*]. We Germans need rye bread. And in the Schume Forest, I can't talk to Odin like I can in Solling.[148] That's why I have serious reservations about moving with you to an area that is alien to us. We both need German, Nordic air.

I'm amazed that I was able to finish this long speech. So, dearest, don't think that you have an advantage over your fiancé because of your cold. Fogginess has clobbered my neck and brain. Hopefully by the time you get this letter we'll both be in better shape. Maybe the evening post will bring me a greeting from my love – then there's no need for any medicine.

With longing love, your Erwin

November 22, 1917, noon
Good morning, my dearest Annamarie! I've only just crawled out of my nest, as I had a real fever, à la Africa, last night – maybe as a punishment because I slandered it so much yesterday. Autumn in Flanders is not my thing, but last year it was even worse. Hopefully I'll be back on my feet soon.

Yesterday evening your dear letter Number 21 from the 17th arrived and it was a real pick-me-up. And you didn't receive a greeting from me that day? This terrible postal system from the field [*Feldpost*]! Every day I wrote a greeting, no matter how short, on one of the pretty Solling homeland cards to win your heart over to my beloved Solling forest.

As soon as I can make firm travel plans, I will let you know in some way in good time. Don't be surprised if you get a telegram with a very official look – others usually don't get through. By the way, you can communicate with me as an official member of the War Relief Office.

As soon as my senses are clear again, I will come to you with a proper letter. Today, I don't even want to reach for my violin – that's saying a lot.

In true love, your Erwin

P.S. The news has just come that there is a ban on going on leave for the next few days. So let's count on the first days of December. If you don't get anything from me for a few days, don't worry! I write every day, at least briefly.

Bavikhove, November 24, 1917, evening
Dearest! Now comes the first ball point pen letter![149]
The point is still a bit stiff, but it's done a few pages, it'll work fine. At the Hubertus Mill I'll be talking to your "Heinrich"[150] so that we can get you some goose quills. These steel pens are not a good invention at all. They may be suitable for court clerks and similar dedications, but they're not fit for lovers and poets – this is why there are no longer any good ones (poets, that is).

Your fiancé has a very bad conscience: this morning the mail left without a word to you – because it left an hour earlier than usual and because I stayed in bed til noon. The stupid fever came back and caused me to shake hard all night. Luckily, yesterday, one, two, three, four heartfelt letters came and were the best medicine. Thanks, my love, and a kiss for each one!

I'll answer each one properly as soon as I can think clearly again. Today I have a mild quinine[151] hangover – I can't even tell "whether north and south fall on the same day," as my friend *Hauptmann* Holz used to joke in Africa.

My only consolation is that you are feeling better.
With best wishes, your Erwin

Bavikhove, November 26, 1917
My dear farm girl!
Today I can quarrel with the postal system [*Feldpost*], which left me for two long days without a greeting from my girl. But there will be something that will come today. I very much hope to hear that you've completely conquered your flu – mine is on the decline. Whenever I sneeze, I call out: Hoplalla Annamareike! – this helps.

Yesterday evening, I read, among other things, Gorch Fock's diary excerpts[152] and came across the phrase: "For a person to be complete, there are two." Do you know what I was thinking? I was thinking that impressions of a general kind, which we receive from outside ourselves, are formed by us into the ideas and images that are very concrete, and which are rooted in our own experience! When we hear something beautiful: good words or noble music, the imagination immediately begins to look for an environment in which to place it.

When your father read us from Gorch Fock, I really listened very carefully, but I was only thinking of you – can a poet wish for a stronger effect?

Whenever I hear the full-throated call of the black woodpecker somewhere, it always conjures up the old noble (where the woodpecker still calls, nature is always noble) Solling Oak Forest before my eyes and, half hidden between fir-scented drifts, there's a white house with a red roof and a green shop, in front of which there's a setter[153] and two long-eared dachshunds – that was the hunting lodge of my brother who died in Russia.

Whenever I hear heroic music, the giants of the alps stand before me and I see the avalanches thundering down the valley.

That's what I'm up to, just as you write to me about yourself. I also prefer to listen to concerts in the morning or in the late afternoon – late in the evening my imagination is usually too exhausted, and listening to Beethoven with the score in my hands is not my thing.

The little flowers that I enclose originate from my last "victory wreath" – flowers are placed around the plate of every lucky warrior in the evening.
See you tomorrow, my dear heart!
Your Erwin

November 27, 1917
Hurrah! My Annamarie is healthy and fresh again! Adhering to your example, your future spouse can report the same to you. Strange that this insidious African fever flies away as quickly as it overtakes me. Today I flew with the squadron again. That was necessary, because things seem to be developing into a big battle here again, and we have to play our part.

Yes, kid,[154] we have to accept that I won't be able to leave before December 6th. Until then, there is an absolute travel ban, inbound and outbound, including my group leader, who I still have to stand in for – by rights he should have come back yesterday.

This postponement and waiting is bad for a heart that longs for his girl – I'll have all the more kisses waiting for me afterwards.
With great love, your Erwin

Homeland [*Heimat*] postcard of the Solling countryside

November 28, 1917
In the spring we want to look for the primroses in the meadows, dearest! We can roam around there all day without encountering annoying people. How I look forward to it! Now I'm waiting for your new picture, but even more for the day when I'll see you again.
Your Erwin

November 29, 1917
Dear heart! Just a quick, heartfelt morning greeting! The squadron is already waiting for me. I'll come to you this evening with a real letter.
Your Erwin

[here a cross appears in the text]

[commentary by the editor of the original volume, Dr. Johannes Werner]: On his second ascent in the afternoon of November 29, 1917, Erwin Böhme achieved his 24th aerial victory behind enemy lines over the Zillebeke Pond, near Ypres.

Soon after this engagement, he was overpowered by an enemy squadron and shot dead over Zonnebeke. The English buried him with military honors in the cemetery near Keerselaarhoek.

The Pour le Mérite that had just been awarded to him lay among the unopened mail on the squadron leader's desk, waiting for his return.

Epilogue

Army High Command, 4th Army
Commander of Aviation [Kommandant der Flieger]
At Army Headquarters, November 30, 1917

At the head of the Bölcke fighter squadron that he commanded, *Leutnant* Böhme fell behind the lines in aerial combat yesterday after 24 aerial victories.

Trained by his friend and master, Bölcke, in the art of superior attack skills, this magnificent, weather-resistant, battle-tested man with rare energy and tenacious drive, undeterred by heavy losses, understood how to win new greatness and victory laurels to add to the old crown of glory of his proud squadron during the Battle of Flanders.

Inseparably linked to the history of the Bölcke Fighter Squadron, the name of *Leutnant* Böhme will remain unforgettable to every pilot of the 4th Army and a shining example of noble, manly virtue.
Wilberg

Commander of Aviation, 4th Army
The Commanding General of the Imperial Air Service
November 30, 1917

I extend my warmest condolences to the squadron on the death of their brave leader. Despite his advanced age he, like the youngest pilots, wrested victory after victory from the enemy at the head of his group of dependable men.

By the highest order of cabinet, which did not even reach the fallen man himself, his and his squadron's merit at the Battle of Flanders was honored by awarding the Pour le Mérite to *Leutnant* Böhme.

In the squadron he was the last of *Hauptmann* Bölcke's close circle of friends. His death renewed the sense of duty to preserve Bölcke's spirit.
The Commanding General of the Imperial Air Service
von Hoeppner

Guard Corps [*Gardekorps*],
General Command
Army Headquarters, November 30, 1917
The air service suffered a new heavy loss with the heroic death of *Leutnant* Böhme. After 24 aerial victories, an enemy shot brought his active life to an end.

The Ypres fighter group stands with the entire air service to mourn at the grave of this fearless and brave officer, to whom the spirit of the great master Bölcke has passed, and whose name and work live on in an exemplary manner in the service.

May the air service never lack these kinds of heroes.
The Commanding General,
Von Böckmann

To Gerhard Böhme
In the field, December 1, 1917
I just received the sad news of your brother's death. One becomes tough and hardened in war, but this case touches my heart very deeply – you know yourself how close your brother was to me as a friend.

On the last afternoon before his death he was still here with me at Avesnes le Sec, my new airfield – he was full of joy about the development of our dear old *Jagdstaffel* Bölcke, which is brought back to its former heights solely by his own merit.

Now they are both united in Valhalla: your glorious brother and his great master, to whom he was the closest of all of us. Please visit me soon, dear *Herr* Böhme, so that we can commemorate this lost brother and friend together.
With deep condolences, your
Manfred *Freiherr* von Richthofen

To Gerhard Böhme
Dessau-Ziebigk, November 29, 1918
Today, on the anniversary of your brother's death, we commemorate the Böhme family, who have always been closely connected to us through the same suffering, and the fallen hero, who of all our son's comrades was the closest to him, and most closely related to his spirt.

How vividly the dear, serious, dignified, upright man stands before my soul! I can still see him on October 31, 1916, at the airfield in Lagnicourt, deeply shaken by the terrible fate that he had to see his friend fall by his side, and I was allowed to speak to him in a comforting way. A year later, we saw him here at the simple memorial service at our son's grave and then had the joy of welcoming him to our home as a guest. With happy excitement he drove from us to Hamburg to see his fiancée – and then just a month later he too gave up his hope-filled life for our fatherland.

Though the pain of our dear dead will never be healed, we should only complain softly as we consider that their heroic deaths saved them from witnessing the betrayal and disgrace that are now about to break our hearts, and because we hope that their sacrifice will not be in vain, but will continue to work as a role model in the hearts of German youth to use all their strength and even give their lives for the fatherland, so it can recover its health and honor. May God allow that![155]

Connected to you by loyalty, yours
Prof. M. Bölcke

Endnotes

1. *Ein ganzer Kerl* – more literally: "A complete guy" – in Anglo-American parlance, "quite a guy" or "quite a fellow."
2. Though his name is usually spelled "Boelcke" for Anglo-American readers, I'll keep to the original spelling "Bölcke" used in Böhme's text.
3. Böhme was shot down behind British lines, and so his effects from his body were kept by soldiers who recovered them at the crash site.
4. It's not clear if it's his fiancée, Annemarie Brüning, or his parents who made the letters available for publication.
5. East Africa was a German colony until the end of World War I. Today this is Tanzania. The Pare Mountains are part of the Kiliminjaro region. Neu-Hornow was a town that has been renamed Shume.
6. The *Gewandhaus* is Leipzig's famous orchestra hall.
7. Footnote in original text: Thus wrote Dr. H. Koenig in his essay dedicated to his old mountaineering companion in the "Alpina" (dated February 15, 1921) of the Swiss Alpine Club.
8. The *Jungfrau* is a mountain in the Swiss Alps.
9. *Hubertusmühle* -- "Hubertus Mill"-- This was a factory complex in Germany, just outside Berlin, that included a saw mill.
10. *Heldenkeller* – a "hero's cellar" or "hero's bunker."
11. This is extraordinary testimony to the violence of industrialized war, that the shells traveling through the air create such commotion as to disrupt the flight of aircraft.
12. A French pusher-type observation aircraft.
13. He's referring to installing an extra machine gun to fire forward in his Albatros C-type two-seater aircraft, which also has a machine gun used by the observer sitting behind the pilot.
14. He's looking forward to flying a Fokker E-type monoplane, which was one of the first aircraft to have a forward firing fixed machine gun, making it Germany's first fighter aircraft.
15. Here Böhme makes a macabre joke about Allied propaganda, which accused the Germans of murdering Belgian children.
16. *"Verfranzt"* – pilot colloquialism "to get lost," making fun of the observer (*"verfranzt"* literally means "to get franzed").
17. This would be the large, twin-engined, Gotha-Ursinus G.I biplane.
18. Johannisthal was the airfield on the southeast outskirts of Berlin.
19. *Soziale Frauenschule der Innere Mission* – This was one of many women's auxiliary organizations in which women volunteered to prepare care packages, help wounded soldiers, and organize other forms of aid. Women's war work was an essential part of home front efforts to secure victory.
20. *"Glühwein"* – wine served hot or warm, often with spices added -- a popular German drink for warding off cold weather, usually in winter.
21. Two weeks before this letter was written, the Battle of Jutland – known by the Germans as the Battle of Skagerrak – took place. Though the sea battle is largely interpreted as indecisive since the Allied blockade of the German navy remained in place, it was celebrated as a victory in Germany since the British lost more ships, including several battlecruisers.
22. "Blue-jacketed friends" -- nickname for British sailors.
23. Though the German fleet couldn't claim a decisive victory, as the British blockade remained in place, they did sink more British capital ships, and here Böhme is joking sarcastically about how British propaganda portrayed the result of the battle as a British victory.
24. *"Königlich Sächsische Schweiz"* – a famous mountain range and park south of Dresden.
25. She's referring to Ludwig Weber, one of Böhme's comrades.
26. He's taking the liberty of addressing her by her first name – a bit bold by the middle-class conventions of the time, where it would have been expected that he address a recent acquaintance by her last name. I'm also translating

the "Honorable (*Verehrtes*) Annamarie" fairly literally. It's similar to "Dear Annamarie" for Anglo-American readers, but a bit more formal, in accordance with German letter-writing culture.

27 It's implied that Annamarie writes letters notifying wives and children about their injured or killed loved ones.

28 Here he uses the word "alagähr," which is not a German word. It seems he might be having fun referring to the Spanish or Latin word "alagar," which can mean "praise" or "flattery," which is needed to acquire luxury goods.

29 Annamarie refers to the piece of paper as a "*Flieger*," making a pun on the words "flier" and "flyer."

30 "Iron arms" – a metaphor for French artillery.

31 These marshes, also known as the Pripet marshes, were in Galicia, which is now the border region between Poland and Ukraine.

32 The Styr and the Stokhid are rivers in present-day Ukraine.

33 Following Annamarie's lead in her previous correspondence about his letter taking off in the wind, he's making a punful joke on the word "*Fliegerchen*" – a "little flier/flyer," or letter.

34 "*Spazierenfliegen*" – literally "pedestrian flying". I translated this to the more colloquial "stroll in the park."

35 He's referring to the Austro-Hungarian Empire, which was allied with Germany.

36 Tyrol was a heavy contested region on the border between Italy and the Austro-Hungarian Empire. Böhme seems to be expressing admiration to the 'brave Tyrolians' who were loyal to the Austrian Empire and fought fiercely against Italy, which aimed to occupy and annex the region.

37 The troops of Austro-Hungary, a multi-ethnic empire, were often divided by language and culture, which created confusion and inefficiency in military performance.

38 Brusilov was one of Russia's few capable generals. He organized a major offensive against the German army a few months before this letter was written.

39 Immelmann and Bölcke were Germany's two most famous aces at this stage in the war. Immelmann's death was considered a huge blow to morale.

40 "*Kaltwasserheilanstalt*" (cold water asylum) and a "*Glashaus*"(greenhouse) are both derogatory terms for a mental institution and a nursing home, respectively. Böhme's rhetoric reflects stigmatization of mental illness, a prejudice that was widespread at the time.

41 Her new street address is named after the sun: "Sonnenau."

42 The "*Rauhen Haus*" was a famous institution in Hamburg where care was provided for disabled children. It was founded by Protestant organizations in the early 19th century.

43 Böhme was thirty-six, relatively old compared to his comrades, at the time of this letter.

44 Originally from Saxony, Böhme is taking a shot at infamous Prussian bureaucracy.

45 "*Jagdstaffel*" literally means "hunting squadron," the German term for "fighter squadron".

46 The U-Deutschland was an unarmed merchant submarine that made a trans-Atlantic trip to the United States in early 1916, before America got in the war, and returned to Bremerhaven on the day that Böhme wrote this short note.

47 As she noted earlier, "Aspirin" is Annamarie's nickname for her difficult boss.

48 He's referring to Hans Kohlschein, a famous artist whose depiction of life at the front was displayed in major German art galleries during the war. Kohlschein did a portrait of Böhme, which is included in this volume.

49 "*Rettungsmedaille*" – a Prussian civilian medal given for saving someone's life.

50 "Aspirin!" has become Erwin and Annamarie's secret word for any person or event that's a pain.

51 Böhme's language has shifted a bit – he's opening his letters with "*Liebes Fräulein*" ("Dear *Fräulein*"), a little bit less formal and more intimate than his previous letters.

52 He's referring to the new Albatros D.I and D.II machines, which arrived at the end of September 1916. Pilots celebrated these as superior to the Halberstadt fighters and the old Fokker monoplanes.

53 "*Heidegrüss*" –literally "pagan greetings," likely related to late September, which is the time for harvest festivals.

54 Andree was a popular atlas publisher in the late 19th and early 20th century.

55 An interesting term here: "*persönliche Verbundenheit*" – very similar to language one would use to describe a relationship with a human.

56 Another interesting term: "*seelische Kontakt*" – could also be translated to "spiritual", or literally "soulful" bond or contact.

57 "*Selig*," in addition to "blessed," could also mean "blissful", or "saved", in a religious sense.

58 "*Alte Kanonen*" – Germans used the term "Kanonen" (literally "guns" or "cannons") for "aces," thus I've used this more literal translation, "old guns," to convey the evocative original language.

59 Hirth was a leading figure in German aviation and aero-engineering before the war. He was also a famous pilot who was wounded in the war.

60 "*Sachsen*" – "Saxony" – the name for the L 9, which was destroyed along with the L 6 in a hangar fire at

Fuhlsbuttel, near Hamburg, on September 16, 1916.

61 *"Couleur"* students – a general term for student fraternities, who were distinguished by their colors worn on their uniforms. They were famous for duels that left scars, which were symbols of masculine prowess.

62 He uses the term, *Corpsgeist*, also known by the French term, *esprit de corps*.

63 *"Förster-Bruder"* – a "woodsman-brother" or "forester-brother."

64 Assistance for families who were dependent on soldiers was one of the growing costs of total war. As human and material losses mounted, the number of war widows and orphans grew exponentially, putting greater pressure on the socio-economic fabric of German society – see Roger Chickering, *Imperial Germany and the Great War*, Third Edition (Cambridge: Cambridge University Press, 2014), especially Chapter 4.

65 Franz Joseph, the emperor of the Austro-Hungarian empire, died on November 21, 1916, aged 86.

66 Complaints from soldiers about authorities on the home front not understanding the stress of life on the front were commonplace. Resentment between those on the combat front and the home front intensified over the course of the war. Again, for the best overview of this, see Roger Chickering's *Imperial Germany and the Great War*, especially Ch. 4.

67 *Jasta* Bölcke went through a quick succession of leaders after Bölcke's death, as Stephen Kirmaier was killed in action on November 22. Franz Walz would take over and remain commander until June 1917.

68 "Vickers" was a generic term used by German pilots for virtually any pusher type aircraft, but Böhme is likely describing an F.E.2 aircraft here.

69 In order for a victory to count, it had to be confirmed by an eyewitness.

70 Ernst von Hoeppner became the commanding general of the German air service (*Luftstreitkräfte*) in October 1916.

71 Food shortages and rationing were taking their toll on the German economy at this stage of the war. With the Allied blockade and a poorly managed war economy, the cold months of 1916-17 were known as the "Turnip Winter" due to the lack of basic foodstuffs.

72 *"Ersatz"* was a term used for "substitute" or "artificial" goods – like *Ersatz*-coffee or *Ersatz*-chocolate. By this stage of the war, with food shortages leading to low-quality, artificial replacements, the term cropped up frequently in dark humor about the harsh effects of the war.

73 Bölcke's *Feldberichte*, which included his reports from the front, was published and became a bestseller in 1917. For an excellent stuy of Bölcke, including translations of his *Feldberichte*, see Lance Bronnenkant's *Oswald Boelcke—The Red Baron's Hero* (Reno: Aeronaut Books, 2018).

74 As an inside joke, he opens with *"Grüss Gott"* – literally "Greetings to God", the traditional greeting in Bavaria, which is where he's writing from in Partenkirchen, near the border with Switzerland.

75 *"Haus Wiggers"* – it was an institution for recuperation of disabled soldiers.

76 "Was hat dieser mörderische Krieg für unsere Volk als Rasse!" The term *"Rasse,"* which suggests humans as a species, is in contrast to the *"Volk,"* which is the equivalent of a nation or a people.

77 This kind of Darwinistic rhetoric about preserving the German nation was frequently found in right-wing, racist, nationalistic thinking at the time. Though Böhme doesn't target any specific "race", similar rhetoric about "degenerates" on the home front infected with antisemitic resentment emerged after the war, and fueled the legend that the German army was not defeated but "stabbed-in-the-back," as many looked for scapegoats for defeat.

78 Böhme's ability to express sarcasm about superfuluous duties to Annamarie demonstrates the intimacy of their relationship.

79 Günter Plüschow was a famous pilot stationed in Tsingtau, a German colony in China, when the war broke out. His exploits traveling secretly through China and then eventually back to Germany in 1916 were recounted in his wartime bestseller, *Flieger von Tsingtau*.

80 The German grading system – 3 or 4 would be average or below average grades.

81 He's referring to a Sopwith 1 ½ Strutter, a two-seater observation aircraft.

82 It's not clear what "triumph" (he uses the word *"Erorberung"* in quotes– a kind of conquest or triumph) he's referring to, as we don't have Annamarie's letter or much context given here. But based on the next lines it seems she may have sacrificed some food, perhaps gave it to someone else.

83 *"Tintenhelden"* – literally "ink heroes." I used the more Anglo-American colloquialism "red tape heroes" for Böhme's criticism of zealous bureaucrats.

84 Böhme comments on *"Orthopäden"* (orthopedist) as a "nice German word" – perhaps it's the Greek origins of the word that makes this seem unfamiliar to him.

85 Field Marshal Paul von Hindenburg became chief of the General Staff of the armed forces in 1916 and, along with Erich Ludendorff, controlled the German government and war economy as a virtual dictatorship.

86 It's interesting that Böhme's gentle digs at the home front

87 "*Komischer Kauz*" – a weird or strange old guy, a kind of oddball old geezer. Böhme is making fun of himself and his age.
88 Moritz was von Richthofen's Great Dane.
89 "*Ulmer Mastiff*" – the German term for a Great Dane.
90 The famous composer Richard Wagner was infatuated with Mathilde Wesendonck, a young poet, and his romantic letters to her led to the break-up of his marriage.
91 "*Frühschoppen*" – a traditional Bavarian Sunday brunch that can include sausage, pretzels and beer or wine.
92 Here Böhme refers to "those whose life revolves around the *Stammtisch*" – "*Stammtisch*" is a German term for a reserved table occupied by regulars at their local bar or pub.
93 Böhme is referring to No. 56 Squadron RFC, which was filled with a number of experienced British pilots and rumored to have been formed to counter Richthofen's *Jasta* 11. Lothar von Richthofen was credited with shooting down Albert Ball, but the circumstances of Ball's death are not entirely clear.
94 Indeed, he uses the word "*Wrack*," the word one would use for a shipwreck, a derelict piece of wreckage.
95 "…zum ersten Mal spüre ich, dass ich im Kriege 'Nerven' bekommen habe." One could translate this to "…I got 'nerves'" or "…I got 'nervous'".
96 "*Kurland*" (Courland), which is present-day western Latvia, was a Baltic border region contested by the Russian and German empires. During the First World War, many 'Baltic Germans' proclaimed themselves to be part of the German empire, but this became Latvia after 1918 and was eventually taken over by the Soviet Union.
97 As mentioned earlier, "Vickers" was a general term used by German pilots for any British pusher type. In this case, Voss likely shot down an F.E.2b.
98 Annamarie is amazed that she's able to get fresh butter from grass-fed cows, as there were shortages of basic food goods by 1917.
99 Here she tries to replicate the German farmer's pronunciation of "Verdun" and the "Aisne."
100 *Plattdeutsch* is the Low German dialect predominant in northern Germany. Though she expresses sentimental admiration, Annamarie's prejudices about these peasants as quaint and primitive also bleeds into her language.
101 Fritz Reuter was a famous novelist who wrote in Low German (*Plattdeutsch*) dialect.
102 The full title of Reuter's novel was *Ut mine Stromtid* (*From My Farming Days*), published in three volumes from 1862-64.
103 Löns was a popular writer and poet who celebrated the natural beauty of Lower Saxony. Löns was killed in action in September 1914.
104 Gustav Frenssen was another popular author who wrote about the spiritual and natural beauty of northern Germany.
105 Löns suffered from numerous physical and mental health problems.
106 Here he's referring to the Battle of Messines in early June 1917.
107 Superstitious beliefs about the number thirteen were widespread.
108 "*Etappenschweine*" ("reserve-lines hogs") refers to soldiers who served behind the front lines, often in administrative or supply jobs, in contrast to the "*Frontschweine*" ("front hogs" – equivalent of "grunts") who served in the front-line combat zones. There were often tensions between the two, especially as "front hogs" were convinced that "reserve-line hogs" got better food and safer conditions.
109 "*Abschusstorte*" –translated literally here as "shoot-down (*Abschuss*) cake (*Torte*)," a strange term that Böhme perhaps made up to describe a celebratory cake eaten on this occasion.
110 The United States just declared war on Germany on April 6, 1917.
111 Kurt Wolff served briefly as commander of *Jagdstaffel* 20, from early May to early July 1917, before he returned to command *Jagdstaffel* 11.
112 Von Richthofen became commander of *Jagdgeschwader* I, which consisted of *Jagdstaffeln* 4, 6, 10 and 11.
113 Manfred von Richthofen suffered a head wound on July 6, 1917, while attacking a formation of British Fe.2d two-seaters. He was likely accidentally struck by fire from one of his own squadron mates who opened fire from behind him while his squadron attacked the British aircraft.
114 In German: "*Flandern*" (Flanders) and "*Wandern*" (wandering or hiking).
115 A "parasol" was a monoplane with its single wing positioned above the fuselage. He was probably referring to Morane-Saulnier type aircraft.
116 The Berlin Secessionists were artists who rebelled against traditional, objective standards of art. They explored impressionistic and expressionist styles, which included dramatic depictions of surreal landscapes, which is likely what Böhme is referring to here.
117 Footnote in original text: Taken from the Major G.P. Neumann edited collected work "In der Luft unbesiegt"(München J.F. Lehmanns Verlag, 1923).

118	Skat was the most popular card game played by German soldiers.
119	Here he means he can approach the English-occupied Belgian coast from the sea.
120	Böhme's joke here is a bit odd, but seems to be disparaging what he sees as the unsportsmanlike nature of "flak" (anti-aircraft fire).
121	This is an odd term – "*kleinen Volk.*" He's commenting on their small stature. Thus his next line about needing to bring on more ballast so the plane would be properly balanced.
122	Kurt Wolff was killed in a dogfight on September 15, 1917, flying one of the new Fokker triplanes.
123	He uses the word "*Konvivium*" here – a Latin word for a group of people who enjoys good food – he's jokingly referring to her group of girlfriends and their parties.
124	Werner Voss was killed on September 23, 1917, flying solo in a famous dogfight against numerous members of No. 56 Squadron RFC.
125	He uses the term "*Kanonen*" – "guns" or "cannons" – German slang for "aces."
126	Von Hoeppner was the commanding general of the German air service.
127	Dessau was Bölcke's hometown and where his parents lived.
128	Böhme refers simply to the "*Riesen*" – nickname for the "giant" bomber aircraft ("*Riesenflugzeug*"), flown by his brother, which tended to suspend long range flights during the winter.
129	Annamarie's letter was from just before they were engaged, when she addressed him with the formal "*Sie*" (you). After they were engaged, they both used the more informal "*Du*" (you), reflecting their more intimate relationship.
130	Böhme's description of the dynamics between white Europeans and Africans is interesting – Africans were treated as inferior by German colonizers in racially segregated societies.
131	Charlemagne rose to the peak of political power as the first holy Roman Emperor in the 9th Century.
132	This is of course the well-known hymn, written by Luther, "Ein feste Burg ist unser Gott" ("A Mighty Fortress is our God"). A loose translation of the second verse: "With our power nothing can be done/we are soon lost/ The right man, whom God himself has chosen, fights for us."
133	Here he's referring to what the Italians called the Battle of Caporetto, which was a catastrophe for the Italian army in the summer and fall of 1917.
134	The "*Erzbergers*" refers to followers of Matthias Erzberger, a member of the German Centre Party who, by 1917, gave a speech in the Reichstag in which he called for a negotiated peace to end the war, which would include compromises with the Allies and limiting territorial ambitions set by the Kaiser. Böhme is clearly critical of this approach to ending the war.
135	He actually writes "*Numero 1*" – French or Spanish – they're now numbering their letters.
136	"*Herzlichst*" – traditionally translated to "sincerely", but the German phrase is more evocatively a reference to the heart, and hence my more literal translation.
137	Johann Wolfgang von Goethe was one of Germany's greatest poets and writers. Letters from his mother, Catharina, were published and popularized.
138	Gorch Fock – a pseudonym for Johann Wilhelm Kinau -- was a popular author who was killed at the Battle of Jutland in 1916.
139	Von der Vogelweide was a Medieval songwriter, well-known for his love songs.
140	By this stage of the war, black market trade and crime was widespread as the German economy began to collapse under growing inflation and shortages – see Roger Chickering, *Imperial Germany and the Great War*, Chapter 5.
141	As mentioned earlier, the Secessionists were a modern art movement known for impressionistic and expressionistic styles, which Böhme is mocking here.
142	Böhme uses the derogatory term "*Polacken,*" reflecting the prejudices that many Germans held toward Poles at the time.
143	Böhme is referring to the popular magazine, *Velhagen und Klasings Monatshefte*, without using its precise title.
144	"*Flanderische Waschküche*" – literally "Flemish laundry room" – slang for thick fog.
145	"*Liebesgaben*" – literally "gifts of love" – translated to the more familiar "care packages."
146	He's referring to the famous Sanke card series of portraits, which were very popular.
147	An 'Afrikaner' is an Afrikaans-speaking individual in South Africa, typically Dutch settlers who colonized Africa.
148	Solling is his home region, with the mountains that he spoke of earlier. Odin was the Norse god.
149	"*Kugelspitzenbrief!*" – "a ball point letter" – the ball point pen was invented at the end of the 19th century, but they were almost never used commercially until the 1930s. The kind of pen he's referring to here, though a novelty, would have been very different from a modern ball point pen.

150 "Heinrich" seems to be one of their geese on the farm.

151 Quinine is an anti-malarial drug, used to fight off fevers.

152 Footnote in original text: Gorch Fock (killed in the Battle of Skagerrak): "Stars over the sea: Diary Excerpts and Poems," from his estate, ed., including a summary of his life, Hamburg, 1917.

153 "*Hühnerhund*" – literally "chicken dog", the German word for a setter.

154 Böhme affectionately called her "*Kind*", colloquially "dear," but more literally "child" or "kid."

155 The narrative here by Bölcke's father, written just a few weeks after Germany's defeat in November 1918, reflects the rhetoric of many conservative Germans who felt betrayed by defeat and resentful of the revolution that would usher in Germany's first democracy, the Weimar Republic, which many Germans, traumatized by the effects of the war, its human losses and political and economic stress, were unwilling to accept.

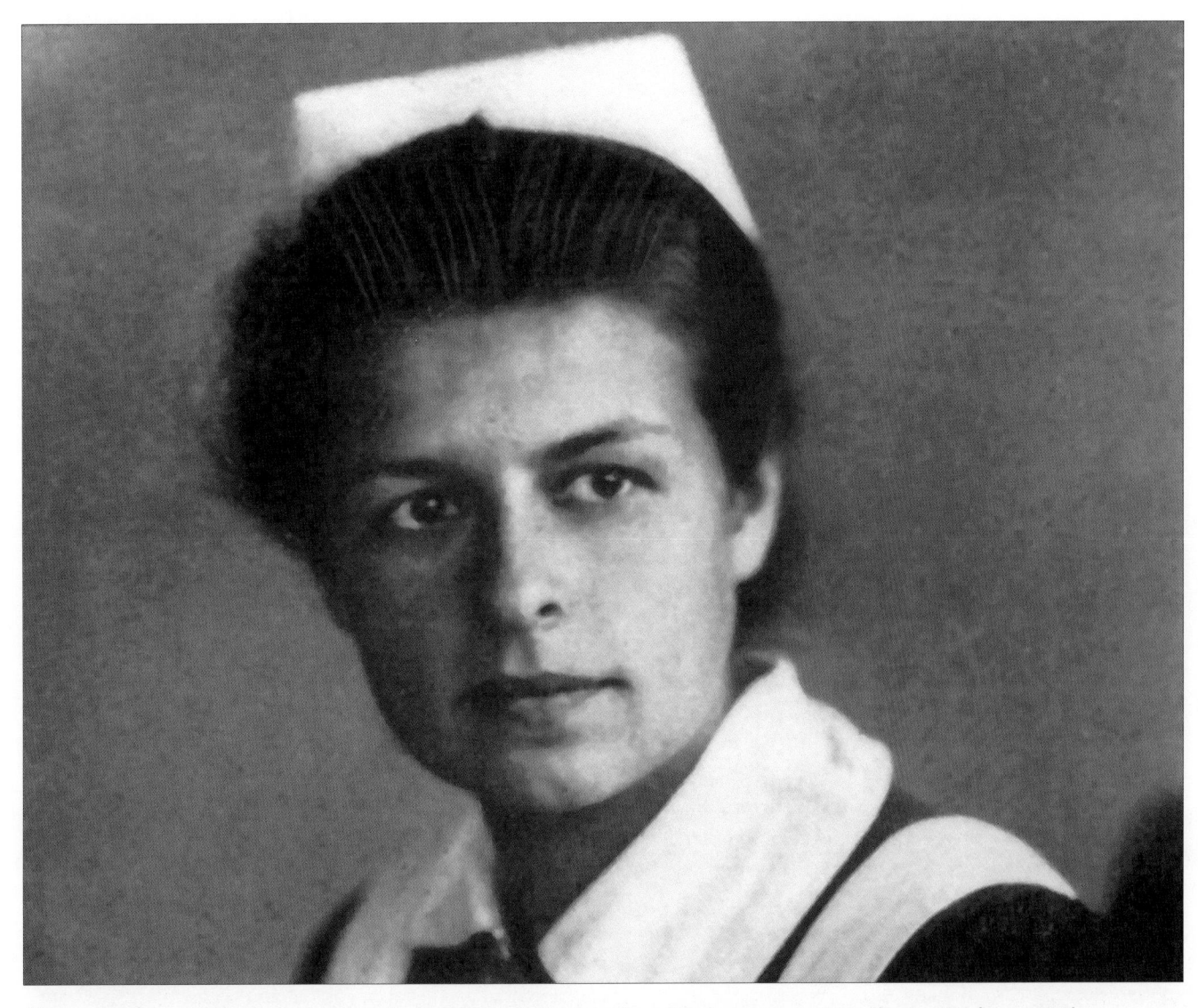

Above: Annamarie Brüning as a nurse in Hamburg, 1917. (courtesy Niedermeyer in *Over the Front* Volume 10:1)

Introduction to Karl Emil Schaefer, *Vom Jäger zum Flieger (From Infantryman to Flier)*

Karl Emil Schaefer's memoir was part of a wave of wartime publications that popularized the image of fighter pilots as mythical heroes. Published by August Scherl, which also released Buddecke's memoir (see volume one of this series), Schaefer's narrative reinforced prevailing conservative ideals of sacrifice and patriotism. Similar to other pilot memoirs, Schaefer's was published shortly after his death, and it begins with a dedication from his father, who expressed his hope that Schaefer would serve as a model for the next generation. Aimed at a younger audience, the memoir promoted nationalism and heroic masculine ideals that typified middle-class values in imperial Germany. Schaefer's narrative was assembled by his father using the son's letters, diaries and notes. Thus, it must be considered to some degree as a construction of how Schaefer's father imagined his son's legacy.

In this context, Schaefer's memoir does not deviate from the formula taken by many of these wartime memoirs that reflected prevailing militaristic ideals. Nevertheless, his individual personality does bleed through. Similar to Buddecke, von Tutschek and Gontermann (translated in volume one in this series), or Böhme (found in this volume), a tension occasionally emerges between the heroic image and a more complex personality just beneath the surface. Schaefer documents the horrifying effects of trench warfare, and his experience being wounded, before his transfer to the flying service. In an interesting example of how men might try to conceal the brutalizing effects of war, he writes what he calls "reassuring" notes to his mother about how everything is fine, while he writes separately to his father about what it was really like to be wounded. On one hand, this suggests a family dynamic that perhaps many men might relate to, but it also hints at the hidden traumatic effects of war that are often sterilized in written accounts.

While Schaefer largely adheres to the rhetoric of heroism and sacrifice typical in wartime memoirs, there's also a tone of melancholy that hangs over the narrative. He reflects on the emotional impact of the loss of friends. He also documents the effects of combat on his psychological condition, using the language of 'nerves' so often found in these sources. Like in many of these memoirs, he

Above: Cover of *Vom Jäger zum Flieger*.

asserts confidently that he can control his nerves, but the increasing stress begins to wear him down.[1] Also similar to other memoirs, Schaefer emphasizes the importance of comradeship in helping him to endure the psychological strain of life at the front. While he finds a sense of belonging with his front-line comrades, he feels increasingly alienated from loved ones on the home front. In May 1915, after recording the horrifying effects of trench warfare, he bitterly criticizes his parents for not writing him more often, suggesting he feels isolated and hopeless. Schaefer continued to rely on emotional support from his parents, but the growing tensions between soldiers and civilians signaled an experiential gap between combat and home fronts.[2]

Schaefer copes with the psychological effects of violence in different ways. This can be seen in his language after he joins the air service, where killing and dying becomes a kind of routine. He writes about getting shot down as though it were some kind of "movie," an out of body experience. Like many pilots, including von Richthofen, his language is also filled with shooting down the enemy as a kind of sport that produces a sense of elation. Such rhetoric suggests a certain psychological distancing from the violence that he both perpetrates and fears. Schaefer seems to be aware of a paradox involved with aerial combat. Pilots like Udet and von Tutschek were conscious of being both drawn to and repelled by it, finding both horror and beauty. Schaefer described air fighting as "nerve-wracking, wild and wonderful."

Perhaps most interesting to readers, Schaefer's memoir offers detailed descriptions of daily life at the front. In addition to the delving into the everyday life of survival and finding creature comforts, he includes his reflections on the terminology of early flight (in his chapter "Introduction to a Flight Manual"). Readers can find here in-depth analysis of words invented by pilots. Schaefer offers a glimpse into the infancy of aviation language as humans tried to distinguish this technology (he insists that laypersons call it "flying" not "driving"!) from all others. This careful description of flying language reveals how challenging it was for pilots to describe their experiences and the sensation of flight to those without experience.

For extensive biographical background on Schaefer, I would recommend Lance Bronnenkant's excellent *Blue Max Airmen*, volume 7, with Aeronaut Books.[3] But some background here might be helpful: Schaefer was originally from Krefeld (also Werner Voss' hometown) in Rhine-Westphalia. He was studying to be an engineer in Paris when the war broke out, and he snuck back to Germany and volunteered for the *Reserve Jäger Regiment Nr. 7*, which he recounts in detail here. In September 1914 he was wounded and hospitalized for several months, returning to the front lines in May 1915. After another brief period of infantry combat, he requested a transfer to the air service and flew with *Kampfgeschwader 2* on the Eastern front in the summer of 1916 before transferring to Manfred von Richthofen's newly formed *Jasta* 11 in February 1917. He was one of Germany's most prolific fighter pilots during the infamous 'Bloody April,' and he shot down twenty-one aircraft during that month. At the end of that month he was given command of *Jasta* 28 and his total number of kills reached 30 before he was shot down and killed on June 5, 1917, in combat with Lt. Thomas Lewis of No. 20 Squadron RFC.

A quick note on language – I've used the spelling "Schaefer" here in this introduction, as this is most familiar to Anglo-American readers. Further, he signed his name as "Schaefer" in the front piece. However, it's spelled "Schäfer" on the cover and in the text of the memoir, and so I stuck with that spelling in the translation. I provided extensive endnotes to clarify Schaefer's use of language and the meanings of German words (some of which are difficult to translate). Endnotes are also provided to help with some context and identify ambiguities or inconsistencies.

Jason Crouthamel

Endnotes

1 The history of psychological stress in war, and changing language and perceptions about it, is extensive. As noted in the intro to Böhme, for an interesting overview of how the language of "nerves" permeated war in the twentieth century, I'd recommend Ben Shephard, *A War of Nerves: Soldiers and Psychiatrists in the Twentieth Century* (Cambridge: Harvard University Press, 2001).

2 Tensions that emerged between home and combat fronts during the First World War have been studied in-depth. For a great book on how the crisis of war affected one particular German city, and shaped experiences of both soldiers and civilians, see Roger Chickering's *The Great War and Urban Life in Germany: Freiburg, 1914–1918* (Cambridge: Cambridge University Press, 2007).

3 See Dr. Lance Bronnenkant, *The Blue Max Airmen: German Airmen Awarded the Pour le Mérite, Volume 7: Bernert, Schaefer, Wolff* (Reno: Aeronaut Books, 2015).

Right: Emil Schaefer in front of his Albatros D.II 1724/16.

From Infantryman to Flier – Diary Entries and Letters

Vom Jäger zum Flieger – Tagebuchblätter und Briefe

by

Leutnant Karl Emil Schaefer

Published by August Scherl Verlag, G.m.b.H., Berlin 1918

Contents

Forward
Hurrah – It starts
To Belgium
At Maubeuge
Infantry [*Jäger*] Patrol on the *Damenweg*
Back in the Field
Soldiers' Burial
A Day in the Lice House
Hurrah "Pilot!"
To Russia!
To the Western Front
Under *Freiherr* [Baron] von Richthofen
Introduction to a Flying Manual
With Richthofen against the English
A Double
The Massacre of Children in Loosbogen
How Richthofen was Shot Down and Wounded
Bad Luck

Foreword

When my son was on vacation and talking about his experiences in the field and his aerial victories, I repeatedly asked him to put his accounts in writing. The answer that I always got was: "I don't have time to write books during a war."

When I was sorting through his estate after my son's death, I noticed that my words had made some impression on him. I found notes on which I recognized that he was thinking about pulling together his experiences and publishing them later, perhaps only after the war.

I have now assembled his letters and diary pages somewhat loosely, but in chronological order one after another. The resulting book best reflects the person of the fighter pilot and his intense love for his fatherland, for which he died. May these pages preserve the memory of those who fell at such a young age in the hearts of the German people and encourage our developing youth to do the same.

Krefeld, early 1918
Emil Schäfer, Sr.

Hurrah – It Starts!

At the beginning of July 1914 I arrived in Paris with the good intention of staying there for a few years. In 1912–1913 I served for a one-year stint with the Hannoverian Infantry [*Jägern*][1] and then went to London. Since I believed that I could finally speak enough English, my father set me up to work, temporarily as a volunteer, with a French silk wholesaler who was a friend of his. I now wanted to stay in Paris for a long time and not come back to Germany any time soon.

Around the 20th [of July], Dad came over to do business, but the Austrian ultimatum to Serbia obstructed his plans. No customer wanted to buy anything – the word "la guerre" [war] was already damaging business. There was hardly anything for us to do either. The staff just lounged around in the break rooms and talked politics. Our old cashier, that short French citizen who is the type of guy who is hard-hearted and works us like reindeer, who is also staunch socialist and pacifist, naturally mentioned the word on our minds. "War?" he said? "There is no more war. We don't want a war. The socialist's dream is a Franco-German understanding. Where would a war supposedly come from? Believe me, *Herr* Schäfer, one might mobilize, but nobody will ever order an attack."

Our fellow traveler in the city, *Monsieur* Octave, was less optimistic, or perhaps more. On the likelihood of war, he had

Above: Frontspiece, Schaefer, *Vom Jäger zum Flieger.*

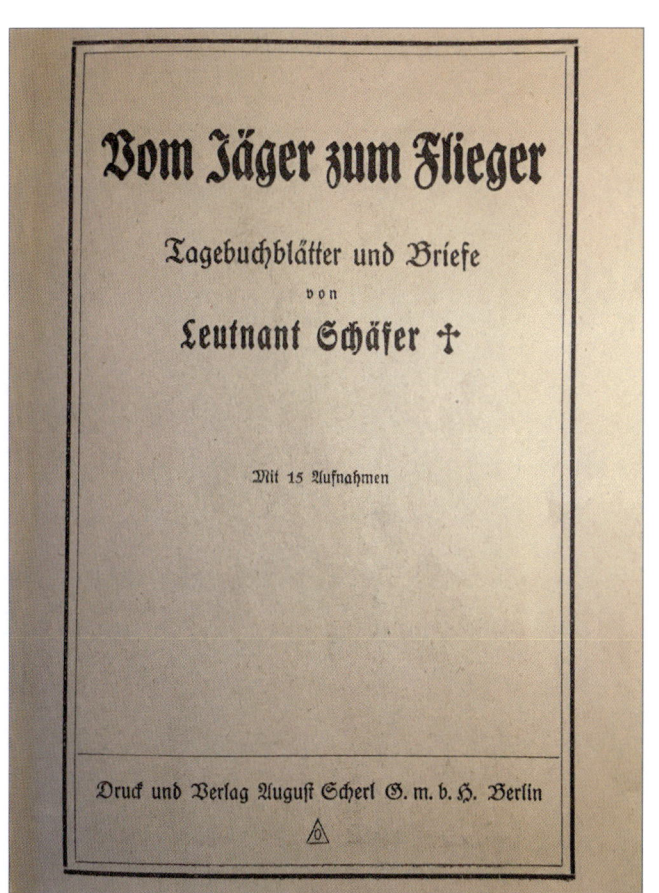

Above: Title page, Schaefer, *Vom Jäger zum Flieger.*

no particular opinion. When I asked what feelings he would have if he were serving as a soldier going off to war, he said: "Feelings? It's not a matter of feelings. You are mobilized, you fight, and when you have the opportunity, you sing." Two young people age 18 and 19, sturdy brats who were both fervent socialists and never missed a party meeting, took part in the frequent anti-war demonstrations every day. They would come to work the next morning after screaming their lungs out, and they had scratched faces.

These two also had really good intentions. "When a war breaks out," they said, "the first thing to do is to mobilize the army and reserves; the untrained troops will come later. Thus we have a week or more to reach Belgium, England or another neutral country." There was only one of them, a little tough guy, who was a peace-loving socialist. "If France does well," he replied to a question I posed, "I will never forget that this is my country." The others teased him because since the ultimatum he had marched four times a day from his apartment to our store just for practice. But he was an exception, a white raven among all the others that I've met.

There was feverish excitement in Paris for a week. Every evening the socialists held prepared and improvised demonstrations, parades and meetings against the war. Of course, there was always a minority demonstrating against the demonstrators. If two crowds of people collided, there was "vive la guerre" and "A bas la guerre," and then they fought.[2] After fighting for a while, the police intervened, arrested the main shouters and then dispersed the rest. Once I saw a couple of German-speaking youngsters being followed by a group of angry people. A platoon of socialists came from the opposite direction, sang while they tried to save the [socialist] party members that were under attack, and then as a fight broke out between Frenchmen versus Frenchmen. The Germans quickly fled through a side street. In another instance there was a huge socialist demonstration in front of the Matin.[3] A strong force of police and civil

Above: Streets of Paris on the outbreak of the war.

guards were ready; counter-demonstrators swarmed around by the thousands, looking forward to the big fight – but in vain. The [socialist] party had issued a secret counter-order and relocated the meeting place to a completely different area. Before the necessary police teams could be thrown into the mix, windows were smashed for several kilometers, street lights were ripped out, police officers and passers-by were injured by shots from revolvers and knife stabbings.

Of course, I enjoyed listening to the crowds and the excited speeches. When the police intervened, I would hide behind the line of people and collect the hats that were left behind on the battlefield. As an angel of peace, I would go out and distribute my goods like the girl from a strange land. There had of course been no lack of paternal exhortations to keep me safe at home or stay in a hotel with Dad. But at the same time, I did not tell my father about the experience that finally made the air too sour for me on such excursions. Otherwise he would have safely locked me up or knocked me down as he did in my, and his, younger days. He would have had plenty of reason to do so. This is how it happened. It was before the Matin. The newest news wasn't yet out, and the public desperately wanted something to do. A young boy thought that I looked like a German – "*sale Allemand*" and "*chien d'Allemand*."[4] I was just barely able to stand up against a house to cover my back, and there were already a few hundred people around me. Five or six meters further to the right was a street corner with a police station and a train for the civil guard. But they couldn't see me, and even if they had been able to, they would have looked in another direction. Running towards them was out of the question. Then a boy of about 16 or 17 asked me if, depending on whether I could manage it with my filthy mouth, I would like to speak French. Instead of just asking him in a few French words to let me pass, I shouted a genuine Parisian curse at him, which ended with the beautiful and clear word "idiot." The result was that, although he was just a tiny, thin guy, he attacked me with an angry roar. I took a step to the side, grabbed his tie, yanked him high up in the air and threw him at his buddies. Thus I got some air. I immediately got the uncommitted members of the audience on my side because of my genuine Parisian answer. It only took a few seconds and I was able to get a few steps around the corner to safety. However, I would not have been missing much even if I were completely beaten up on this occasion. In any case, when I made my way towards the quiet cloister in the Cité Trévise, I swore never to take part in these boulevard demonstrations again. Besides, I shouldn't have any more opportunity to do so.

The following day, Tuesday July 28, I was sitting in front of a Café on the Poissonière Boulevard, when a noticeable unrest arose on the Italian Boulevard. There was running, screaming, a crowd of people, and above it all the screaming of the street vendors: "*Paris-Midi*."[5] When this newspaper comes out comes out at noon, all you have to do is close your eyes to see the prairies of North America. This shrill "*Paris-Midi*" sounds like the battle-cry of the Sioux Indians. This time there seems to be something very special going on. The public literally snatches the newspapers out of the boys' hands. I saw quite a number of such rascals appear one after the other. But nobody came up to me. One vendor appeared in a back alley barely a hundred yards away with a fresh pack of newspapers. When I got halfway there it was sold out. The same thing happened at a newspaper kiosk, to which a cyclist had just brought a huge pile of newspapers that were still wet from the press machine. In my mind's eye I can still see the woman literally wiping her fingers, blackened by the press ink, on her apron, while she lets out a substantial curse. From the comments and fragments of conversations that I hear from passers-by, I figured out what was going on, and eventually I managed to get hold of a fresh paper. "In Germany the order for general mobilization is to be given."

Ten minutes later I got out of a car in front of the German consulate. A quiet, elegant hall, marble stairs, a friendly official who knows nothing, a few rooms, a large, semi-dark hall with a crowd of young and old faces, in which I can see the same question that drove men here.

Is it starting? Before we could even think about this, the counsel general was already here. "Gentlemen, I can imagine what brought you here. I have not yet received any official confirmation of the news, and I am not expecting it today. By the way, as things stand, I am strongly convinced that we will have mobilization in a few days. Anyone staying here in France for pleasure, or to learn the language, or if you are temporarily on business, should leave immediately. If you leave straight away, you make one more space free on the train for those who have to stay until the last minute.

I will be available in the next room if you have any special requests. Good morning, gentlemen." Short, clear, military-like. I could have almost ripped myself to shreds when he left. When I got back on the street in the midday heat, I could not think clearly. The word "war" kept surfacing in my mind. There was a big mess in my head. I climbed into a car or truck, stepping on a couple of children and whatever else was in my way. I went to the nearest swimming pool and cooled off my head. Then I went to my old man, and the next morning we packed, bought tickets, reserved our seats and the like. At lunchtime there was a small farewell meal in the hotel for we two Germans, a Dutchman, and a few Frenchmen, among whom there was a salesman, his wife, his first employee, the owner of the hotel, and my boss with his son who returned from Germany that morning. Nobody wanted to believe that the war was happening. We were laughed at when we left, but drank to hope that we would meet again, and then we parted in the best friendship.

When we got to the Gare du Nord [train station], President Poincaré had just returned from his trip to Russia, and so we could hardly get through to the train on time. Aha, we said – and we said it many times on this trip. The railways are taken over by the military. Aha, soldiers on the train on heading off for the front. Military transports, artillery, infantry are all heading in the direction of the Belgian border…In Belgium, factories stood still because people were already mobilizing for the borders. And so gradually we "aha'd" all the way to Krefeld [Germany], where Mama stood waiting at the train station. When she saw me in front of her so unexpectedly, I could have summed up her train of thought in a single "Aha." However, she put it a little differently and took me in her arms. Mothers of heroes don't cry – well, it's easy to get something in your eye when there's a draft.

To Belgium

Bückeburg, August 2 [1914]
Dear Parents!
My efforts in Goslar to move out with my old battalion were in vain. My draft order is for the 7th Army Corps, and nothing is to be done against the order to go to war. So I have been here since yesterday. Tomorrow I will be in uniform and in six to eight days I will be sent to the battalion that is supposed to leave this evening. Nobody knows where we are going. Some claim the Belgian border—Aachen.
                    Your soldier

Liège, August 21, 1914
Dear parents!
Today I hope to have time to write a little more than this short postcard from the front (Feldpostkarte). After a few really heavy marches – one of them was almost 50 kilometers – we marched into Liège late in the evening the day before yesterday, and then yesterday evening, while the company was quartered in the old citadel, I went with six people to a small suburban train station to protect the railway. The place was totally deserted. No train had come through for twelve hours. There are no available personnel left. The local population seems really anxious and is as friendly as possible – but nevertheless, one has to be careful, of course. My revolver is always in hand. We've taken over the guard room here from the 39th and set up the place to be quite comfortable for the time being. It was a bare waiting room – the only furnishings were half a dozen long benches. This has all been transformed so now there are two long tables with benches along the wall with windows facing the train platform – this is our living and dining room.

At the end of the table, there's a bench that divides the small room in which my desk is located. This and the writing implements, stamps, lights and tools are gathered from a few nearby offices. On the opposite wall there are four mattresses for anyone who wants to relax. Nearby our rifles[6] lie neatly on a bench, the back of which has been converted into a rifle rack with a few strokes of an axe and some nails. Finally, in one corner is the washroom, made up of two washcloths with a bench, a towel and a mirror, all things that we kidnapped from the rooms of the boss and the second-in-command of the train station. So the place that was bare and uncomfortable this morning now makes a quite comfortable domestic impression, and the only problem is that we will probably be relieved in a few hours.

So, as you can see, I am doing extremely well. I can deal with the stress, which up to now hasn't been too difficult, and I feel very good. At the citadel there is a detachment of reserve field artillery, and I met *Leutnant* Hasenklever, with whom I had a long conversation. Max Escher is also there as a deputy sergeant. In Liège I met the young Herbst, who is a lieutenant in the 39th, and I met at the commandant's office the tall guy from Krefeld, Oellers, of the 10th *Jägern*, who had been captured in a night battle and was released again during the counterattack. He said that *Leutnant* Wendeburg was dead and *Leutnant* Reuder was very badly wounded.

Now I hope that this message will actually get into your hands.
                A thousand greetings to you and my sisters.
                Your loyal son and brother

Liège, August 23, 1914
To my dear ones at home!

We have been at Liège for a week now. Eight days ago we marched for 50 kilometers and, though we were really tired, we still enthusiastically sang the "Watch on the Rhine" and "Oh Germany in High Honor" on the way to the conquered fortress. Forts shot up by gunfire, burned villages, destroyed rail lines, fallen trees by the side of the road, streets torn open – this all gave us a sense of the war. We now thought that we would find a picture of horrific desolation in Liège itself. We were disappointed. The core of the city, whose surroundings and outer part are so devastated, has remained almost completely intact. All large public buildings are intact and have been put into military service. The palace of justice, a wonderful gothic building, houses the headquarters in which there are clerks, orderlies, and adjutants receiving orders from incoming troops and transports, medical officers, aviation officers, motor and carriage transport, Belgian police and who knows what else. Around fifty automobiles, mostly confiscated private Belgian vehicles, are made available to headquarters. There is a similar operation in the large Lambert square. All entrances are blocked by barriers against civil traffic. On the square itself there is a company of armored infantry protecting the area, and to the right and left of the entrance to the headquarters there are two guns, and a row of machine guns on the large balcony. Around the square, cafes and hotels are still in full operation. I was able to go outside for two hours today, and so I looked at the all the hustle and bustle and for the first time since going into the field I ate a civilized lunch. It cost me almost a decade's wages, but it was nice. We even drank champagne; when will that happen again? It will hopefully be more French when it does. The way to Paris goes from here to Reims, the capital of Champagne, so it better be cheap.

I wrote to you while on guard at the train station. Unfortunately, it lasted only one day and one night. I passed the next night in the citadel; magnificent view, romantic surroundings, but a powder keg. Our chief was delighted when we arrived downstairs. Now the mines have been flooded and the artillery shells and other ammunition supplies have been removed. But our company, the guard for the battalion commander, received the order that was intended for the division commanders' elite troops, the *Jägern* [infantrymen]. After the active troops had withdrawn on our arrival, only the 13th Reserve Division remained here as a temporary garrison. Now, of course, everything is restored, and we're actually superfluous. In the first few days, however, things were a bit busy, and only one company, ours, was available to protect all 15 nearby bridges. For two days and one night I stood with a dozen men at the Pont de la Boverie, a bridge over the Maas river which connects a long road from Verviers, and I also had six motor boats from the commandant's office guarding a few hundred yards away. For every double-duty watch I had only three men, while six are mandatory. There was little peace and quiet, and I took watch myself if a guy got too tired. Later on we got a few reinforcements, first from the *Jäger* and then [regular] infantry, and now I have quite a sizable armed force of 30 men, three corporals and one non-commissioned officer. I joined his people on one end of the bridge, of course the less dangerous one. He had his guard room inside a furniture moving van. I myself rest with my *Jägern* at the other end of a cleared-out newspaper kiosk, which I got from an abandoned house with spring mattresses, quilts and kitchen furniture. The bridgehead is ingeniously set-up. The blast holes, which the Belgians had already dug, were expanded, reinforced with a wall made out of paving stones, and protected against stone splinters with earth and sandbags. That probably sounds a bit ridiculous for a bridge defense watch in the middle of a town that has been in our hands for two weeks and is occupied by several army corps, and yet it is necessary. The guards standing a little further away by the boats were initially fired upon every night, but it was not possible to determine where the shots came from. In general, there was regular shooting during the first few nights we were here, if not in my area, then at the neighboring bridge or in another street nearby.

On the first night that the division was in Liège, some of our men were lured into hidden spots and murdered. One in a stable, another in a bar, one was pulled from the Meuse river, and so on – thirteen men.

Three days ago at 11:00 at night a company marched along the river bank, just opposite my post. I wanted to test my new spyglass for its nighttime suitability, so I looked at the company, the head of which was just turning a corner, when revolver shots came from the houses around us and the entire first group of men fell. Three were dead. The captain was wounded. We were bitterly angry at this wretched gang that blends in around us during the day and then ambushes our men at night with pistols and shotguns. During the incident, which took place on the other river bank, barely 300 meters away from us, we were of course ready to fight in our self-made fort, and when on our side of the river a couple of guys took advantage of the turmoil to shoot out the windows of a company standing over there, it was time for us to get them. Unfortunately, I was not allowed to leave my bridgehead. I would have loved to have planted a sidearm and cleared out their nest. Unfortunately, we had to leave that to the infantry reserve. It was a very eventful night and our 39th cleaned up a lot. The restless elements are mainly workers-for-hire, who of course don't have any employment, as well as Belgian soldiers in civilian clothes, of whom a few thousand are said to be in Liège, and a few hundred Russian students, among whom there are anarchists and nihilists of the worst kind. After the measures taken last night, however, it has now been quiet and the reasonable and better elements, who are in the majority, breathe a sigh of relief. It was unavoidable that sometimes an innocent man had to suffer with the guilty. It's night again now. It seems like things want to stay calm. So far only one defender has fallen. Trouble will probably come from a nervous guard. After all, it's no wonder if people start seeing ghosts.

I'm sitting here in the boardroom of "*L'air liquid,*" a secret club, on the first floor with first-class electrical lighting. The Meuse river rushes under the balcony, the doors of which are wide open to let in the beautiful air. Even from up here I can hear my people on guard snoring. Only the double watch tromps slowly up and down and interrupts the silence of the night. It's not always that quiet, by the way. Transports and troops come through here at virtually any time of the day or night, moving down the big staging road that leads over my bridge from Berviers. It is quite astonishing to see what columns of motor cars, medical units, field bakeries, butchers and ammunition troops get through here. I just keep wondering where you get all these countless cars, horses and wagons. It's outrageous. Prisoners and wounded also come by, as does the field post [*Feldpost*]. The *Feldpost*, the damn… leisurely, dear lovely *Feldpost!* It was first given out yesterday, on the 21st, and then again today. Do our things to you also arrive so late? The second delivery was still possible, but the first was a bit bad. I gave my first long letter to a Düsseldorf painter who drove it straight home [to Germany] in the car. Has it arrived? I also hope to be able to give this message to a car that is going at least as far as Aachen. As a guard on the bridge in the reserve area, I simply stop cars that look a bit "suspicious," ask where they're from and where they are going to, and then the whole thing is done. "In war you have to be a bit bold [*frech*], then anything goes," as our captain says.

Since we've been in Liège, we have only been living on canned Belgian fish. Everyone on bridge watch has received a large pot of beer and a bucket of jam [*Gelée*] from the company – strange what the captain finds. The good Belgians, for understandable reasons, have a guilty conscience. They have often run away and left everything behind. So it is clear that you have to take what you need in terms of food. Wherever people were sensible and quietly stayed at home, not a single piece of sausage was taken that was not paid for. Well, I've got to close now – I've never written such long letters in my life.

A thousand greetings from your front soldier

Liège, August 26, 1914
Tomorrow morning we'll be ready to be replaced. What next? Hopefully we'll get to engage with the enemy soon. Service on the police force no longer really suits me. There's already been action at Müllhausen, Metz and Brussels. Namur has fallen. When will we finally get there?

Sunday, August 30, 1914
In Lobbes, behind Namur, the order is given to us by the army: "Maubeuge is surrounded. We are awaiting ammunition. Great victory in St. Quentin, army cavalry 60 miles north of Paris, East Prussia is safe from the enemy!" That helps us get through a tough march of over 5 kilometers. The "Wacht am Rein" sounds fresh and joyful for the first time in 5 days.

We are supposed to take part in the siege of Maubeuge. The local rural population tell us how on Friday the French came through here with music blaring, singing and making noise – on Sunday, the same Frenchmen came through here, but in flight, chased by German bombs.

At Maubeuge

Tuesday, Sept 1, 1914
We are certain to be attacking Maubeuge. We march off with gear for the attack – 12 kilometers. Now we are lying scattered about in a field and listening to the artillery battle. We're not attacking yet. The Krupp 42cm guns are astonishing. They are standing so far back that you can't hear the explosion of their firing up here. But it crashes through the air as if an express train were coming from far away, and it vibrates past us with a rattle. In front of us in the fort a column of smoke and dust rises 100 meters or more. Debris and chunks of wood fly through the air with the dull, jarring smash of the explosion. Scattered bits of fragments and stones fly all the way to land near us, about 1000 meters away. The effect is frightening. After seven shots, five of which strike right in the middle, the unfortunate fort is silenced. Even without the binoculars you can see what kinds of eerie holes are torn into the walls. Nobody within it can be still alive. Fire now comes only from the distant field artillery positions.

Saturday September 5, 1914
We spend the night in a firing position on the edge of the village. I am laying low with three *Jägern* 100 meters in front of this position in an unfinished trench.

It's a very atmospheric evening. The broad plain is calm and peaceful in the uncertain moonlight. On the left, the silhouette of Fort Leveau, black and threatening against the evening sky, emerges with angular contours, which are interrupted by the irregular cracks and fissures of the holes made by artillery fire. Straight ahead the horizon is blood red – over there, behind a ridge, Maubege is burning.

Flares fly through the air. Patrols against the enemy come and go, and during the whole night the concert, to which one has gradually become so accustomed, is not silent. The bright, sharp bang of firing guns close in behind us. The howling, drunken, hissing projectiles fly over us, and you can hear the dull echo of shells striking and pounding in the distance. Only now and then does a French shell find its way to us. There's a rapid burst over us. One hardly looks around for it and we don't even duck at all.

September 6, 1914
It is night time, almost 2am. The first platoon is on watch in the field, and besides the sentries, I'm probably the only one on watch, although nobody is supposed to be asleep. For the entire day the battalion followed the battle and slowly advanced towards the still intact *Fort des Arts* behind the right wing of the brigade. We saw the first dead lying on the road. Wounded people passed by, here and there a stray bullet whistled past, and shrapnel burst over us. But we haven't done anything, while the Pioneers of the 39th and 24th have struggled and suffered hard. Now we are here on watch, 600 meters from a village in front of us, which was fought over this afternoon. It was in our hands once before, so that the battle equipment of the 39th was already there. But at around noon the fort fell to us again after falling under a prolonged artillery attack.

Dispersed infantrymen and the Pioneers camp in the field while they're on watch. I have just tucked some of them, who arrived dead tired and immediately began to fall asleep on the bare earth, into straw beds. *Leutnant* von R. also crawled into his sleeping bag. He hasn't slept in three days. I wake up instead of him and write these lines in the bright light of the moon. The general situation doesn't seem to be rosy. The 28th Brigade is the right flank of the troops advancing on the Maubeuge. In the left is the 14th Division, and on the right only a weak cavalry unit. There is a whole lot of artillery here. Fortunately, a mounted machine gun unit arrived yesterday. This is how we get into the fortress, which is very important because of the railway. But if the enemy recognizes our weakness or even has the heart to attack us on the flank, then we can get down and dirty, and the fate of the brigade may then depend on us *Jäger* digging in somewhere on this flank and defending it down to the last man until help comes or until the brigade has been properly dislodged from it. If they don't take us on, then tomorrow we'll probably attack the last fort between us and Maubeuge.

Either way, there is hard work tomorrow. I'm pleased to say that I'm looking forward to it as a long-awaited thrill. I'm overcome with real anticipation.

Monday, Sept 7, 1914
After laying about and waiting all morning, it finally starts at noon, and around 3am the first lines of defense are set up. A small, shot-up fort falls without an infantry battle, then it starts against Fort Leveau, a very strong defensive position that was only shelled by the artillery during the morning. When we came up to a ridge, we see the Rothofen plain from over here. Do they want to attack? That pitiful, whimpering sound we hear can't be the signal to attack. We get the order to stop firing: Maubeuge wants to surrender.

Our first platoon comes to Fort Leveau to serve as guards for a prison. We guard 2,800 prisoners with 80 men. Later, two more companies of infantry come to reinforce them. I work as an interpreter and a casino board member and set up a very cozy officers' quarters with the *Jägern*.

Tuesday, September 8, 1914
A French officer has to leave me his horse, a captured dragoon gives his spurs, and I do a nice ride around the area. It looks great! A shot crashed through the roof of the main building to the basement and shook the whole place up. In the rooms, people are still sitting with their hands in their pockets, arms crossed, heads propped in their hands, on benches and chairs or at the table, etc. – dead – killed by the violence of the explosion. In one corridor there are 6 or 7 dead in a pile with their skulls cracked and their brains blown out – gruesome. A shot hit in the courtyard and dug a hole into the earth into which I put 280 prisoners. People make amphitheater-style seats with their shovels and seat themselves comfortably. At lunchtime there are two groups that ride in first, then after I take ride on my horse, the entire first platoon is moved to Maubeuge. We arrive there at 2am and watch the prisoners move past the gate. It's 52,000 men whom we and 18,000 other German troops forced to surrender, and the march past lasts until 12:00 the next night.

The solemn parade of prisoners must come to an end. In no time, we just fall asleep on the pavement with our guns. Then at 1:00 in the morning the order comes to march. From the very start, we can hardly drag ourselves because we're so tired. We have to go to Frammery – a twenty-kilometer march.

Thursday, September 10, 1914
The original order to march to Antwerp is changed. We're going to Paris – hurray! And the luggage will be driven there.

We walk on clouds and cover 28 kilometers without batting an eyelid. We get there early and there's a straw bed that's not too bad in Bachant.

Friday, September 11, 1914
As a result of the make-shift rooftop and pleasant, cool weather, the mood is downright high-spirited, and even an uninterrupted march of 3 ½ hours does not spoil our humor. A bottle of Rotspon wine and butter make a delightful breakfast. Unfortunately, the neighborhood is just as miserable as the weather. Etreaux is the name of this nest – it was 30 kilometers away from us.

Saturday, Sept. 12, 1914
It was supposed to be a short march. They announced 12-15 kilometers and then a day of rest. We should be bitterly disappointed. An army command ordered: somewhere up at the front things are not going smoothly, so we're needed and have to get on with it. We waltz off, at first in the pouring rain and on awful roads. The weather is improving, but after we've covered 44 kilometers by 7am and have a three hour break, it starts storming again. During the night we march 32 kilometers without a significant break. After an hour's rest, it's another ten kilometers to go and then we'll get warm food. Then we're off again, and at around 3pm on Sunday, September 13, after almost exactly 100 kilometers of marching, we get ready for battle.

In the morning crossed Laon, which because of its magnificent location made us all extremely happy. The area is already becoming fertile to the south. Tomatoes, grapes and other fruits thrive in abundance. If we get into action right away, we'll have all the rest that we need. People in Germany will probably be talking about our march for quite a while.

Infantry [*Jäger*] Patrol on the *Damenweg*[7]
Because of our forced march, we're a bit of a weak link on the last stretch of defense in front of an enemy army. We must dig in and hold on until reinforcements from the XV Army Corps and the Bavarians arrive.

The headquarters of our 13th Reserve Division is at the Malval farmstead on the *Damenweg*. In front of it on the hill is the 13th, 39th and 56th Regiments (II. Battalion). Behind them are the infantry at their disposal. There's heavy artillery down in the valley, field artillery up on the hill. Somewhere to our right is the 3rd Corps, and to our left is the 14th Division, which is grouped in a similar way to ours in front of the village of Khamis.

At about 3am, our link to the 14th Division was cut off. *Leutnant* von Reden, myself and ten infantrymen were sent to reestablish contact. We move along ahead of the firing line. Everything had run away, been salvaged or fled. A few guns stand alone in the field and are still firing. *Oberleutnant* Hasenklever was at his post manning two of them, buried in the shrapnel fire. Here and there were a few infantrymen, holding their positions alone.

At last, happier images arrive. Heavy artillery stands behind the mountains and roars. Machine gun unit 7 is doing well, holding the positions abandoned by the infantry alone for eight hours. A combination of infantry on bicycles and guard regiments 3, 4, 5, 7, 9 and 10 prepare to attack again. A battalion made up of all sorts of infantry sets out to attack a village. In short time, it moves forward. We finally find the staff of the 14th Division in Chamois and go home satisfied. It will be late night before we arrive back in Malval. We report our tasks to the division and battalion commander. *Leutnant* von Reden must still receive orders given to the division, and then we eat at the field kitchen and move into a hayloft. In the morning we get a mild taste of sound sleep while the company has been digging all night. Well, we had been running all afternoon and into the night while under heavy infantry and artillery fire, and so we deserved this little bit of sleep.

On Wednesday, September 16th, early in the morning, it was announced that men from the 56th who were wounded on Monday were still lying on a hill near our position. I volunteer with a group of infantrymen, and a few stretcher bearers, to go help. We succeed in evacuating only about six wounded because, when we walk around out in the open, we get hit by shrapnel fire. The infantrymen retreat to safety. I'm lying too far out on the slope facing the enemy with *Gefreiten* [Corporal] Bittner, so we duck into a crevasse where we receive gunfire if we stick our noses too far out. *Musketier* [Infantry Private] Molkowski lies there wounded. I give him something to eat and drink. I detect some enemy artillery, make a sketch of the view, and Bittner crawls back with it. When the English stop shooting at me, I crawl further and load up a few more poor devils; finally, I even find four who have dragged themselves to a straw hayloft. Klingen and Hamacher, coming from the company commander with food and three cigars to pick me up and help if necessary, find me here at 3 in the afternoon. The four in the hayloft get their food and we start our way back.

I'm given a great reception by the company. On this same evening, Bittner gets an iron cross that's really just superfluous, or so I'm promised. I also share my observations with the artillery and they treat me with gratitude by giving me some red wine. The night passes fairly quietly, with only one alarm.

September 17, 1914
At 4am I lead an officer of the foot-artillery forward. The mortar battery hits the spot that we wanted to occupy, so Lieutenant Gottschalk goes back to correct them. I find a good observation post on a small straw roof and from there I measure how to place fire on two English batteries. I go back and report. The batteries are shot to pieces. Everyone is really friendly to me. The captain says that I should get a good night's sleep.

Friday, September 18, 1914
Once again, there's no chance of getting any sleep. At 1am I was sent out to see if the artillery batteries that I had reported on, and which the artillery shelled yesterday, are still there. I take Bittner and Klingen back with me, as well as Richards; on the way we tell ourselves that the job is almost impossible to carry out in the middle of the night. I'm convinced that we can't get it done, but I'm determined that we do it on our own. We'll go down to the little hayloft. When we get there I get rid of all the extra things that are somehow a hindrance, such as my coat, belt, Shako[8] etc, and set off in just my cap and tunic, armed only with binoculars and a revolver. Nothing can be seen from this far away through haze. I just have to get closer, but my friends will only get in my way. Like a pathfinder, I move forward and get to the flank of the cannons, about 2,000-3,000 meters away. Thank God the English turned on a light for the guard on their artillery piece. I can see that they're still there – my job is done. I get back to my patrol, which is worried as I was gone for more than two hours. Enemy patrols run past without seeing us. I clearly hear the words [in English] "Those bloody Germans don't have no courage to come to this side at this time of the night." The unsuspecting angel!

At 5am we arrive with our report and they immediately send us over to the division. I make an appointment with the leader of the foot artillery [*Fussartillerie*]. At 8am two salvos are fired, and I observe and report the position of those shots as precisely as possible. At just the right time, my patrol and I arrive back at my hayloft. The salvos fall inaccurately, but these are easily corrected. Meanwhile, *Leutnant* Zingel and three members of his artillery crew have laid a telephone line to the edge of the forest, through which I can make corrections and agree on a few simple signals for fire control. After three attempts I place fire on the enemy batteries. One of them tries to drive me away – by waving a signal I succeed in landing a full salvo into their emplacement, shooting them to a pulp. While waving I'm occasionally shot at by patrols, but I'm able to stay safe. *Leutnant* Zingel climbs up to me and in the next moment falls, striking his head. Bittner runs over to help me bandage him. We make the seriously wounded man as comfortable as possible and wait for the stretcher bearers. Suddenly, the hayloft is hit with shrapnel, which bangs around like it's going through an empty wind instrument. Bittner is shot through the foot and then catches a piece of shrapnel in his back – dead. The *Leutnant* gets up as if he wanted to run away, but then he stays down. I jump about twenty paces to the side in a shell hole where I spend the most humiliating half hour of my life. Earth, iron, stones, lead, fire and smoke spray all around me. When it gets quieter, I crawl back to the edge of the forest and can see from there that the hayloft is burning. Despite the gunfire, I run. Bittner is already charred, and *Leutnant* Zingel is gone. I search the whole area with the medics, and we are constantly being fired upon by English patrols. The *Leutnant* is and remains missing. After two hours I give up the search and return alone. I had already sent the patrol away to get help. I report to the field and foot artillery observation posts and then return to the company. The captain receives me like a son who was thought to be dead. In the evening I'll be promoted to *Vizefeldwebel* [vice sergeant]. The following order goes out from the battalion: As of today, *Oberjäger* [non-commissioned officer in the *Jäger*] Schäfer, officer in training with the 1st company, will be promoted to *Vizefeldwebel*. The time spent in the battalion previously is counted as training. I would like to express my special appreciation to *Vizefeldwebel* Schäfer for his conduct. – signed von der Groeben

They care for me in such a touching way. I get dry stockings and shoes, wine and cognac to get me on my feet again, and there's plenty of food, as good as one can get here. At the same time mail came with all kinds of letters and lovely packages. I got a royal night's sleep and was only woken up once by an alarm. It was very nice.

Courtecon, September 20, 1914
Dear Mom!
A thousand sincere best wishes for your birthday. I even have something as a birthday gift, though it's mainly for my own benefit; but I think you will be happy. Above all, despite this long break from writing, which was caused by the very intense military events, I am still well and healthy. The day before yesterday I was promoted to *Vizefeldwebel* two weeks ahead of schedule, and after they gave me considerable praise for my conduct I received the iron cross yesterday. The reason was that I had a few successful patrols that resulted in, among other things, the destruction of two English batteries that had done us a lot of damage. Many greetings to all of you and an extra birthday kiss for Mom, from your front soldier.
Sunday, September 26, 1914
This is a date that I will not likely forget any time soon.

The alarm was sounded at 2:30am, and a quarter of an hour later, we were already on our way to Chivy, which had been occupied by the 4th Company for two days. At the Southeast entrance, the battalion gets ready to attack and at 5am we begin to climb up the hill. It was a tedious, steep climb. I was the captain's personal orderly with Klingen and another *Jäger*. The moment our front line pokes its nose over the hill, it's met with savage gunfire. A fierce battle unfolds, in which we seem to be all alone. The captain sends me to fetch machine guns which, when I arrive with them, are to be used on the right flank, since there is a risk of fire from that side. When I try to get a suitable position, I'll be able to take a shot. Two *Jäger* drag me down the steep embankment to a reasonably level spot where I can lie down, though I'm at a crooked angle. When the English advance, I manage to bring a bit more misfortune to the shot-up cottage with a couple of energetic sounds that bring their weakening line to a standstill again. But now our own shrapnel fire is starting to hit us. If only our guys could shoot just a hundred meters further away – just a hundred meters! I'll have to send the enemy back myself. So I lie down real quietly and prepare for a visit from Tommy Atkins.[9] So he doesn't get my most valuable possessions if he empties my neck pouch, I take out my iron cross and hide it in my chest pocket. It's too bad that my revolver is in my coat and the coat slipped off when I fell and ended up on the embankment. Well, so far no one came near me because shrapnel was bursting right and left, above and below me, one bomb after another. The thought that I might get hit never crossed my find. Actually, I had a close shave,[10] and I had the reassuring conviction that this was enough. Not far from me there was a guy with a severe stomach wound who was groaning horribly. Twenty paces away was *Leutnant* Salge, who kept screaming for help in the most pathetic tones. He was then picked up by a couple of retreating *Jäger* using a strip of canvas.

I was all alone, as far as I could see in my narrow field of vision, with one dying man and nine or ten dead. So I made myself as comfortable as possible on a few rolled-up tents and coats and tried to regain strength through sleep for what was to come. After I dozed for a few hours – it was probably a partial blackout from the loss of blood and pain – I woke up because the gunfire had stopped. The first thing that I saw were two *Jäger* stretcher-bearers climbing out of the steep slope with a stretcher. They crawled through the wire obstacle, and then they reached me and said with all sincerity that they had come specifically to get me. That made me really happy. On the entire way back, despite all the bumps and jolts, I sang, cracked jokes and smoked cigarettes. We got to our defensive lines, which had been built into a position that was really safe from attacks, but which also gave us little opportunity to counterattack ourselves. The battle seemed to have come to a standstill at around 10am. They featured a motley mixture of *Jägern*, infantry [*Pionieren*], 56ers [*56ern*], machine-gun crews, as well as a few from the 39th. *Jäger* officers and elite *Jäger* no longer seemed to exist at all in this sector. I was taken to a room made out of thatched straw in a little house near Chivy and bandaged up by *Vizefeldwebel* Hilgemann with a makeshift but very nice bandage. Three more people came in and then we were left alone. The noise of battle could be heard – first far away, then nearby. The statements of men who arrived here always contradicted each other. The village was continuously bombarded by heavy and light artillery with shrapnel and shells. It was a witches' cauldron [*Hexenkessel*] in which helpless, wounded men like us did not feel very comfortable.

Around dusk the fighting and the bombardment seemed to stop, and I fell asleep, somewhat exhausted. At 11am we were woken up and picked up by patrols from the medical company. The one-and-a-half hour march on the stretcher was a bonus pleasure. After a brief greeting with the battalion doctor in Courtecon, we continued in the car to Colligis, where I was happy to inhale the chloroform fumes on the operating table, as this finally relieved me of the awareness of my pain and my horrible situation. That was September 26, 1914.

Colligis, September 29, 1914

Postcard I

Dear parents!
After I was able to write about many good things to you on the last postcard, today there is the happy news that I will soon be with you again, as I am no longer needed here. They're using a big bruise on my left leg as an excuse to send me home for a couple of weeks. At the next opportunity for transportation I'm supposed to go to Aachen. For the time being I'm lying in an officer's hospital in a French village, in excellent care.

 Many greetings,
 Emil

Postcard II

Dear Dad!
Card Number I is to reassure Mom. But I want to tell you in detail about my injury, so you don't have any unnecessary concerns.

I received a shot through my left thigh from an infantry rifle. The bone is completely shot through, and the exit wound is about 4 or 5 centimeters through my flesh; but this is in excellent condition. The attending physician, the head of a clinic in Paderborn, promised that there will be a complete but lengthy recovery. I'm in a cast and able to go home, but I'll stay here for a few more days because I'm with a friend in an almost completely empty hospital. Please prepare a big smooth bed and lots of cushy pillows, as I have to rest for about six to eight weeks. My overall condition is excellent.

        Greetings,
        Emil

Wiesbaden, February 8, 1915
Kaiser Wilhelm Hospital
Dear parents!
First let me answer Dad's questions: in the house and on short walkabouts and good weather I walk without a cane; otherwise, I take it with me, especially in the dark and when it's icy, etc. I should get rid of it as often as possible so that all my muscles, including the lateral flexing muscles, are forced to work. The lateral cramps in my leg will pass, and the irritation in my nerves from last week has completely disappeared again. When I was injured, the doctors gave me a year to heal, but I think I can do it in half that time.

        Best greetings to you all,
        Your soldier

Back in the Field!

On the *Damenweg*, April 29, 1915
It's a strange feeling to see this area again after such a long time. When the weather is wonderful, I sit out on the land. The ridge on which we got stuck and brought the big retreat to a standstill stretches out in front of me. Out on the top of the ridge is Malval-Ferme, the headquarters of the division, and Courtecon is on the slope. That's where the field kitchens were at that time, where you could always get some good grub. It was there that I "liberated" the wine from the catering officer just as everyone was fleeing from a few falling shells. People really have a good memory! They rubbed my nose in my misdeed first thing when they saw me.

It's so quiet and peaceful here in the valley. The sun is shining brightly, the flowers are fragrant, the guys are singing, laughing and joking while shoveling. Over there on a large meadow a company from the 39th is exercising. Officers and horse grooms [*Pferdeburschen*] move through the column, rifle shots ring out happily from the target stand of the honorable riflemen's guild of Colligis Valley; these are the cyclists at the shooting school. From Pancy comes the sounds of an infantry marching band, and a tiny 18-year old *Leutnant* sneaks through the bushes to shoot crows or magpies. Far, far from the direction of Reims and on the other side of Soissons, the thunder of artillery rolls without interruption; gunfire flares up from time to time near our positions. Only when a pilot flies over the valley every now and then do the guns wake up in our area. One shot follows another, and in the blue sky the little white clouds burst with sharp explosions. They encircle the black bird and follow it through all its curves and evasions far behind the lines, when it finally turns to flee.

When I came forward through the protective trench for the first time, I was really surprised. I had expected an improvement in our trench position, but such a luxury amazed me.

Down at the bottom of the valley, where the little Paradise Forest is, there is a magnificent football field, and two well-trained teams were just about in the best position to replace the men who had been wounded after experiencing minimum combat losses recently. At the edge of the forest with a view of the valley and on the playground, numerous small terraces and gardens have been built, all very nicely made out of birchwood and decorated with shrubs and flowers. Every plant has a mysterious hole in one corner. When the air is bad, you can go into the bombproof bedrooms and living rooms. In the background there are three magnificent buildings that emit wonderful scents. These are the kitchens for the officers and men.

The entrance to the company commander's apartment is protected by a couple of threatening artillery pieces, apparently mortars. On closer inspection, there are a couple of heavy caliber guns mounted on roughly-carved carriages.

A little further away is the cemetery. The fencing is birchwood, and there are beautiful wooden crosses, carved or soldered, or stones with beautifully carved inscriptions. All graves have carefully tended blooming flowers. This is a serious note in an otherwise breezy, yet not fragrant, picture of the camp. But otherwise it's friendly and peaceful, sitting there in the spring-fresh, green forest.

Hunting lodge Malepartus, May 2, 1915
Dearly beloved family!
It's just not nice of you to cut me out completely for a week and a half, not to write, and not to send me anything, although you know that I went to the world war with only one shirt. I've already received letters, postcards and parcels from hundreds of other people. Only my family keeps silent. Is that what one calls a sense of family, or do you think that that it's just no longer worth it? The Prussians don't shoot so fast, and the French are even slower. I'm only sending you this letter to give you a stark example of your wickedness.

        Your soldier

Pentecost 1915 in the trenches!
By a clear, silvery moonlight we set out for Chivy on the Saturday night before Pentecost. The area was brightly lit. The little village peaked out peacefully from the green of the orchards and poplars. Nightingales sang, and only seldom did a shot crack the whip to break the stillness. We go through the village and into the trenches. Everyone goes quietly to their previously designated post. With a handshake and a few whispered words one wishes the relieved comrade a "happy Pentecost." I walk slowly around the position, from the upper right, where a French grave lies by the apple trees in front of our barbed wire, and over on the left, where a cannon emplacement peers out menacingly from the undergrowth on the ruins of a barn.

A sharp watch is kept everywhere. Between every two bends in the trench defenses a guard stands on the front parapet or sits on top of the rear defenses and peers sharply ahead into the terrain. His machine gun stands ready under his cover. One press of the finger and it hurls its deadly hail across. Standing over the boxes of hand grenades are reliable people ready to throw the dangerous apples at attacking enemies. The flare pistols lie peacefully in their holsters today, as they aren't needed against the light of the moon.

Our senior infantrymen [*Oberjäger*] quietly patrol back and forth in their sectors. Up above on the molehill is the same picture of peace. You're tempted to sit down and talk to a good comrade or maybe a lovely girl. You want to adore the moon and hum a little song. One could do just about anything – if it weren't for the fact that over there, barely 30 to 40 meters away, the evil enemy waits.

The view from the forest fire down below into the landscape is delightfully beautiful. Chicy, Beaulne, Verneuil are in the valley. In the background is Beaulne Hill, on top of which the wire emplacements of the 39th glow like cobwebs in the moonlight, and far beyond that in the valley one just barely sees the cannons, the mortars, and the howitzers, which the sleep tonight under light, colorful bushes in honor of Pentecost.

The night will soon give way to a radiantly beautiful day: Pentecost Sunday. – There is no bowl of May punch down here, but we do have an extra good Pentecost roast and a glass of champagne. The dark cellar is freshly whitewashed, there's birch branches wherever you look, on the table, on the walls, in the door. Even the entrance to our cave has turned into a small grove of May poles. There's no holding back today. Not a single hand does any work. We stretch out completely in the green grass in the shade of a few late-blooming apple trees, smoke one of the good cigars that the prince's mother recently donated, and really laze about. Thus we celebrated Pentecost down here, and our opponents apparently did the same, because not a single shot was fired on our position. Only towards evening did the usual alarm break this idyll. At 7pm the telephone rang: "This is the battalion. Message from the division! This afternoon at 4pm, Italy declared war on Austria." "Finally!" somebody said. There a slight pause until another spoke thoughtfully, as if with deep pity: "What a beating those poor, stupid pigs will get." That's the mood of the whole battalion –contemptuous pitying and with a slight shrug of the shoulders, content in our own strength. "Well, if it has to be!" The news was rather well-suited to raising our spirits. Then we drank alcohol again for dinner. Before that, however, we went out and listened to three roaring hurrahs echoing through the mild night air from Troyon to the Beaulne Hill in response to the declaration of war. It may well have been like that on the whole front. And, in answer, the French uselessly squeezed off thousands of rounds….

Soldiers' Burial

June 22, 1915
It was quite solemn. First the bands of the *Jäger* and the 39th, then the honor guard, a platoon of the training company, then three coffins, a dozen officers, a group of stretcher bearers, the half platoon of volunteers from the 1st company as representatives of the first line, then the 39th and 57th came from the front line. Finally, all sorts of people arrived from the surrounding villages.

The small village churchyard of Colligis could hardly contain the procession. The funeral, like all military funerals, was short and solemn. There were a few simple, poignant words from the field chaplain, some soft but crisp commands, somber horn music, a brief prayer, and then: "Group turn to the right – march! Straight forward!" "I had a comrade",[11] and when we turned back onto Dorfstrasse, the horns were singing again: "Rejoice in your life, because the lamp is still burning."

Who knows which of us they're carrying out tomorrow?

Afterwards, the path led me past the churchyard again, and I went in to visit old friends.

It looks odd, this small village cemetery. A few weathered monuments and hereditary burials of long-established residents take up most of the space. The graves of the poor with iron crosses and funeral wreaths made of wire and iron seem almost comical. One says "to our sister [in French]" in crooked enamel letters on a rusted wire frame, small and neglected. Most of the decorations had already fallen down or been removed. Over there a piece of old French jewelry hangs on a German soldier's grave with the inscription: "My dear brother." Thus someone had to lay their brother to rest in someone else's soil and didn't know that he could do something nice for him by taking the ornaments away from where he originally found them.

There are behind us bright, carved sandstones with inscriptions like, "He died a hero's death – He sacrificed himself for the fatherland – Here he rests from the fight – our dear comrade," etc.

In the middle of the churchyard, against the background of a dark cypress tree, stands a white stone on a high pedestal that bears the words visible below an iron cross: "To the brave warriors, the 36th reserve field hospital. Behind it, the high, crumbling wall has been breached, and one steps into the soldiers' cemetery, where a high pyramid of gray stones proclaims

"Germany's heroes," to whom this piece of earth is dedicated. The view sweeps far beyond the simple birch gate over the battlefields of September 13, 14, 15 and the 26th[12] to the ridge behind which the opponents have been facing each other for nine long months. It also shows that the stonemasons of Chamouille cannot keep up with the grim reaper who sits on the molehill over there and waits for the unwary, from Braye to Troyon. He ripped men from the 39th from the right flank, the 57th from the left flank, *Jäger* from the middle, and now they're all lying here together under fresh mounds and wilting flowers. Here in the new part the gravestones set up by the troops, comrades or relatives are still few and far between. Here there are more simple black wooden crosses lined up in a row, which bear the names in white of men who are buried together in groups of three or four. Poor fellows, they were happy to get away with just a big scar and had survived their last uncomfortable transport bravely only to come here and be shown their last resting place. Now they lie here, soldiers to the last, in groups, lined up as if by rank, facing the enemy. Here is the place of honor of the 13th Reserve Division. It's not good to linger too long in a cemetery like this. It darkens the mood. Let's go!

On the way back I found a cross whose inscription read: Field Hospital 33. Grave Number 19. The following died a heroic death for their fatherland: *Unteroffizier* [non-commisssioned officer] Reese, 1st Company Reserve Infantry Regiment 56; Reservist Feldmann, 3rd Battalion, Reserve Field Artillery Regiment 13; Infantryman Fritz Meyer, 2nd F.K, 7th Infantry Battalion ; and Frenman Henri Manaud, 5th Company, 57 Line Regiment.

August 19, 1915
Today I wanted to go to the French emplacements, but after a few steps I found the way back – my secret path – blocked by barbed wire. They were even still working on it. But tomorrow I'll have a look and see if there's anything I can do about it. I won't let my path be blocked without doing something.

August 20, 1915
I went out again today with four brave war volunteers. In the past, you could crawl up into their positions, in the forest, past the wire entanglements up on the hill, until you finally looked out over the concealed sappers' position [*Sappenkopf*], in which a French double guard stood. But now the gentlemen had pulled up the wire and, rather than have an emplacement, they built a fence, which was taller than a man, and there were between four or six rows of these, with the last ones being worked on yesterday. First, we crawled on a trail at the edge of the forest, where we could move silently. We worked our way through three fences and put a man at the end for safety. Then we set about dismantling the foremost fence. After working for about ten minutes, an enemy detachment came down through the forest to this position. We were ready to fire. Unfortunately, they couldn't get close, but they began to work on the sappers' position, building a new fence at the back while we were busy tearing down the first one. We had finished taking care of business and had unwound three nice big rolls of the best barbed wire when a couple of shots were fired and our sentry crawled back. From fifteen paces the Frenchmen shot at him three times. He had not lain still, as I had ordered him, but had crept forward to pick up a white enamel sign that he had noticed in the grass. So now they saw us. The guard shot into the bushes from time to time so there wasn't much point in staying here. So we went home, gave their wire to our wire to our infantry officer Rückert, and I sent the little sign to the battalion, from where it was passed on to be entered into the register for debts to America. The sign was inscribed: [in English] "Made by American Steel & Wire Company, New York, U.S.A. Genuine Glidden Barbed Wire."[13]

Damenweg, September 2, 1915
We're back in Chivy. Our trenches are still in bad shape. A lot of extra frills have been added, but the important work has not been completed. Our anger is mainly directed at a clever gentleman who gave us a role as guests for six weeks and in four weeks managed to become the most unpopular comrade in the battalion. Here is his last deed, which I have put into verse due to its shamefulness:

> Hadn't been in the trenches for a long time.
> We used to clean them up with brooms
> To make them look better,
> And we'd make everything real nice.

But there I was astonished to find the trench was full,
    Covered over in white sand.

The shooting platforms are filled up,
Mostly overgrown with grass.
A part is even sunken in,
If possible, you have to lean out over the edge
But the trench, I have to say with tears of joy,
Is covered up with white sand.

Some of the front and rear defenses have fallen apart,
And all the beautiful sandbags
Are mostly gone, many torn,
Many bulwarks have collapsed.
But the trench, I joyfully report,
Is strewn with white sand.

Defensive positions, formerly so proud,
Built of earth and wire and wood,
They sank to the left and fell to the right,
And no longer protect us if we are engaged in combat.
But for today the whole trench is
Covered with white sand.

The mighty heavy barricade
With a gun turret in the front
Smashing the wire obstacles,
Is now disintegrated, washed away by the rain.
But the trench, no, I'm happy to say,
It's covered in white sand.

Someone smeared the signposts with ink pens for strangers,
And since we had rainy weather the other day,
I can't read them anymore….
But the trench – and this is totally justified –
Is now strewn with white sand.

These comic verses sparked lively applause from everyone. The major also laughed heartily at it, and when the noble guest leaves in two days, no tears will be shed for him.

Now we have one more day to mess around here. Then the program is complete: one day in Collingis, three days in Asfeld, ten days in Colligis, ten days in Bovelle Ferme, ten days in paradise – will it be handled calmly and without problems, or will we finally experience a change? On the whole Western front the 7th Reserve Corps is the only unit that has not changed its position since September 1914. Don't we get to engage the enemy at all? The successes in the East only makes us more bitter about our meager and exhausting service here. Are we still soldiers? Being an officer here is about nothing more than being a better foreman and occasional night watchman over a bunch of earthworks. As a free bonus, one person gets killed here and there, one today, two tomorrow, then none the next week, then six in one fell swoop. They say that whoever has the stronger nerves wins this war. I'd rather just attack the molehill than do this night watchman duty. During our relief, I almost got hit in the noggin.[14] I had climbed out of the ditch at dusk when they started flinging heavy bombs at us. Before I could get back in our tunnel, one fell only ten meters away from me. I threw myself on the ground and was extremely tense for half a second, then the beast exploded all over me. I lay under a blanket of dirt. The glass on my wristwatch was broken, but nothing else. I was glad when we finally arrived safely in Colligis.

A Day in the Lice House

At 4:30am, 240 lost *Jäger* gathered around me, and I had to lead them into the lice house [*Lausoleum*]. At 5am we proudly boarded our extra train, with people in the goods car [*Güterwagen*], me as "*Herr* Transport Leader" in the service compartment of the baggage wagon. As a result of a broken track, we made loud clanking noises for two kilometers. The sanitation facilities are great. A sprawling sugar factory is being renovated. The company advances from the unloading ramp along a fenced path to the gun storage area, where straps and buckles are removed and the guns are put away. Airy, clean hallways take people inside until it's their turn. On the walls are splendid pictures explaining the whole procedure in advance, which saves the *Unteroffizier* [non-commissioned officer] on duty a lot of talking. Two counters are set up at a narrow exit, and each man receives six identification tags on his left and two coat hangers on his right. At the next counter he hands in his valuables, which are packed into a bundle. The first tag is attached to these and they're put on a large rack. Then you go into a big hall, where behind wide counters infantrymen and garrison duty guards stand and rummage through our backpacks. Bold signs help them complete their work: "Everything that can withstand heat and steam is handed over here: laundry, woolens, tent accessories, cleaning supplies; but no leather!" Everything is put in a clean bag that is tied to a dog tag. Delicate items are delivered to the next stand: cartridges, biscuits, sausages, chocolate, bits of iron, bread, etc. From here you go across the yard to the undressing room. There the man hangs all his clothes and laundry on one hanger, all his leather goods, tarpaulin, bread bags and the like on the other and then he hands over his last offerings at the second counter. Now he has nothing – except lice. A large sign warns: "Don't hang leather goods on the clothes hangers. This is what leather goods look like that were accidentally left inside clothes." Below is a sad collection: a wrinkled little child's helmet with an unnaturally normal spike on top,[15] a child-size glove that was once a giant driver's glove, a shriveled piece of leather, a wallet from the olden days, and similar beautiful things.

But now back to my *Jägern*, who are still only human. They are quickly checked for scabies and other nice side effects of lice and then they are sent to wander around in the bathroom. It is a wide room with sixteen huge iron vats painted white. In each of these are gigantic bathtubs, a good dozen little people are kicking around under the incessantly flowing rain from warm showers, the steam from which envelops everything in a dense white mist. This part of the event involves a lot of screaming and laughing, strange clapping and squeaking, and it seems to be enjoyed by those involved as much as the spectators. Anyone who has been cleaned up enough, or, it would be better to say, whoever has splashed around enough, then hikes up the stairs and is given a towel and a woolen blanket at a counter and can now roast themselves in a friendly rest by the heater, play a game of *Schinkenklopfen*,[16] rest or annoy those who are trying to rest, according to preference and inclination. Three quarters of an hour later the shop is opened in the neighboring room. Our cleaned-up stuff is set up on wide counters, the numbers are called out, and the colorfully covered Romans, Indians and Arabs become *Jäger* again.[17] Whoever is finished walks into a half open hall, gets their luggage ready, buckles their rifle slings and enjoys the excellent beer brewed in the neighboring village, with a rich selection of hot sausages and other delicacies offered by the canteen to men who have not exactly been spoiled after twelve months of life in the trenches. After I inspect this routine, with peace of mind I put my men in the trustworthy hands of the doctor, who answered to the unusual name "Meyer" and desired a bath himself. I was taken to a comfortably warm undressing room. The elegantly white bathrooms, impeccably working service room, and very comfortably furnished rest room would have been respectable for any big city pool. Those who want to forget their troubles will find comfortable linen suits and can lounge or read in deck chairs, write letters at a magnificent oak table, or play games in a cozy corner.

After a good breakfast, I allowed myself to look around behind the scenes again. I inspected the machine-generated heating system, the water supply, etc. When we were done, thank God the broken track was repaired, and we had a new train, which was radiant and clean, that brought our cohort back to the battalion in a few hours.

Hurrah "Pilot!"

Köslin, January 31, 1916
Dear Mom!
Unfortunately, a vacation didn't work out. I had to follow orders right away.

So it was a success. Despite my shortened leg I am a well-trained pilot at the military flight school in Köslin. I am looking forward to a 4-5 month peaceful training period as a pilot and I will only be able to go on leave after I've passed my first pilot's exam. You can put your mind to rest about the supposed dangers of flying. Think about it – our facility here has already trained 500 pilots and sent them into the field, and not a single one broke his neck during his apprenticeship. The leader of the command, who has been a flight officer for 4½ years and has earned an iron cross first class, has never in his life seen a crash. For my apprenticeship, I had the good fortune to be assigned to the airfield's chief pilot, a very capable and infinitely careful aviator, so every safety precaution is being taken and you don't need to worry. I'm out of the trenches once and for all – that's certainly worth something too.

With greetings and kisses,
Your son[18]

North Sea Resort Leba, Baltic Sea, June 30, 1916
Postcard to the family of Emil Schäfer, Krefeld

Sincere greetings from a very nice crash,
Your pilot

Obedient Greetings,
Lampel
Leutnant and airplane pilot,
This time as Franz[19]

To Russia!

On July 28, 1916, I went into the field for the third time. In the evening, after a festive farewell to the *Flieger-Ersatzabteilung* 8 [pilot replacement unit 8], we departed Graudenz and then went to Warsaw via Thorn. On July 30 in the morning we went further to Brest in Lithuania, and then took a transport further to Kovel[20] arriving at 9pm at *Kampfgeschwader* 2 [Combat Wing 2] of the Supreme Army Command. Assigned to *Staffel* 8 [squadron 8] led by leader *Rittmeister* Ziegler is my observer *Freiherr* von Grote, who his currently a deputy adjutant. There are military aircraft of the *Bayerischen Flugzeugwerke* (Albatros) including the LDD (light biplane) C674[21] with a 150hp Benz engine. In addition, the *Geschwader* has other machines including a *Staffel* of G-type Rumplers, each with two 160hp or 220hp engines, and four *Staffel* of LDD with 160hp Mercedes engines. The squadron lives on the train. Each officer has two compartments of a sleeping car with the bulkhead removed; chow is in the dining car. The commander, up til now *Hauptmann* [captain] Kastner, a nobody so far, lives in the lounge car with his adjutant. There are rooms for business, cash management, an office, non-commissioned officer and crew quarters, workshops, depots – everything is housed in these train cars, so we set up seven long trains. Our train with living quarters is on a specially built side track four kilometers from Kovel between Russian barracks that are being used as military hospitals. Our tents are on the parade ground, and each aircraft has its own tent with all the accessories in it, where the service crews, non-commissioned officers, and mechanics sleep. In addition to us, there are three Prussians, one Bavarian, and one Austrian unit here, with a total of 80 machines – sometimes it's a terrific operation. Our activities are lively, but not very strenuous and not very dangerous. While the field units are conducting reconnaissance near and far with binoculars and cameras, we are fighting the Russian rear communications in the area of the Stokhod positions. This includes Zaresze, Trojanowka, the Kovel-Maniewize railway and the Kovel-Luck railway. Sokul and Waldläger are in this area, and that's been the scope of our activities for the last ten days. We usually take off in squadrons at intervals of half an hour to two hours, sometimes twice or even three times per day. We fly to our destination as a close group, and then pepper troop columns, warehouses and battery positions with machine gun fire. Occasionally we roast captive balloons or shoot down some Russian planes. So far only six have dared to show themselves. Five of them have been shot down by us, one was shot down by the B.U.K.[22]

The first general attack of the Russians was delayed for 24-36 hours because the half-Asian troops, when they were suddenly attacked from low altitude with bombs and machine guns, simply let themselves fall apart.[23] Six regiments each reported up to 250 dead and wounded. A machine gun detachment of 24 guns was attacked by an airplane at 5am. The driver reported in the evening that he had only been able to assemble six vehicles after that. A Turkestan brigade[24] sent into action in Sokul came under bomb and machine-gun fire from our squadron of G-type aircraft and was completely shattered. Early on August 10 we put Maniewieze train station out of action for 24 hours. The Russian General Command did not have a telephone connection that evening. All in all, we have had successes here that nobody expected and the A.O.K [*Armeeoberkommando* – Army High Command] in Linsingen is tremendously gratefully to us because we saved them at least once, if not more often.

Overall it's going well for us. Flights rarely last more than two hours, and there are almost no flights after midday; we live nicely, rather simply, though a bit expensive, and we're well-cared for. The life of comradeship in *Staffel* 8 is very nice. Unfortunately, we don't have much to hunt. Hopefully there are good chickens somewhere. Ducks are plentiful and are eaten in high quantities every few days, as are crabs, and there are tons of those, only 4 Pfennigs each.

A few days ago I was in Brest in Lithuania; it took seven hours by train to go there and back. I flew it in 2 ¼ hours. It was a desolate, shot-up nest. One feels sorry for the poor fellows who have been sitting around at the airport for months. I'm glad that I was able to get out.

Today I want to tell you about a normal flight against Russia, so that you can see how things are here. For example the squadron order for the evening reads: "*Staffel* 8 and 9 get ready for take-off at 6am, bombing flight, fuel for two hours. The rest of the squadron should go at 6:30am. At 5:45am I'll speak to the gentleman pilots on the landing area [*Landekreuz*]." At quarter to five in the morning my assistant appears and throws me out of bed. Half an hour later I appear in my oldest and dirtiest outfit in the dining car, where the gentlemen from *Staffel* 8 and 9 gradually arrive. Coffee, milk, sugar, bread and occasionally a little bit of butter is delivered. Eggs and canned liver sausage are usually available for a fee, and almost everyone has their own pot of jelly, marmalade, butter or sausage. This first breakfast is very hearty and plentiful, because if you're lucky, it has to last for the whole day. The conversation is very lively and, in addition to discussions about the previous evening's experiences in the "town" of Kovel,[25] the main questions are: How will the weather be, where are we going, with which squadron does the old stag (our commander H. von D.) fly, and whether the Russians might show up today? Five minutes before 5:45 everyone races to the cars and away we go to the airfield.

The airfield is a former parade ground for Russian infantry. A row of airplane tents, probably a kilometer long, now stands on its east side. When we arrive, the squadron is already set up outside, the engines are running and then get carefully covered up again. My C 674, as the leading plane, is on the right flank. The crew stands at attention and the non-commissioned officer reports: "C 674 is ready for take-off! The motor is making 1300 revolutions, it's loaded with eight bombs and 12 liters of petrol." I walk around my good old barge,[26] checking a turnbuckle here and there, a screw connection, then I get geared up. Schubert helps me into my flight pants, buttons, buckles. He ties up the various fasteners, puts on my crash helmet, head cover, goggles and gloves. In short, he gets me dressed. Immediately afterwards, he says: "The *Rittmeister* [cavalry captain] is coming." The small group of men on the landing field comes over. The squadron leader, *Rittmeister* Ziegler, gathers the six crews around him and issues the order: "Take-off at 6am, assemble at 6:30 at 2000 meters above Dubowa. Depart at 6:35am. *Staffel* 9 takes off from Wolka at the same time. Destination: Maniewicze railway station. Climb upwards in a right-hand turn. Approach from about 5 kilometers to the right of the runway. Approach target from 80 degrees to the NW (Wind from the NW at 2500 meters). Turn SW after the descent. Assemble at 2000 meters above Lake Mielnica."

Grote, my observer, and I walk to our crate, and as we walk we enjoy the pretty scene of *Staffel* 9 coming from their distant tents to the take-off area. They are led along like racehorses on the paddock. A mechanic holds on to each wing, sometimes pulling in long sweeps, sometimes carefully pushing supporting, or carrying the plane over bumps. The stocky little biplanes look smart, and the warm brown color of their hulls gleam in the bright morning sun, rivaling the white of their wings and the glint and spray from their propellers. Our Rumplers look like long-legged racehorses compared to the Hungarian Jucker, and the big AEG,[27] which is just taking off to test its propeller and roars like a heavy Belgian horse with its 440hp. We somewhat clumsily mount our machines, and while Grote settles in, storing maps and instruments, as well as testing the machine gun and bomb launcher, I get our ride going. "Switch-off!" screams the mechanic. "Switch-off!" I answer – that is, the ignition is turned off. Now the man spins the propeller several times and yells: "Clear!" I yell out the warning again, "Clear!" – with a push on the ignition lever and quick turn of the crank, the engine starts rattling and rattling. Woe to the poor guy who cries out: "Clear!" He's ignored and remains within reach of the propeller. My skin, arm and legs vibrate. Now the motor must be allowed to warm up. Meanwhile, I have to look carefully at my levers and valves, test the controls, and then bundle up and buckle up. Once the motor is warm, I quickly run it at full throttle, then throttle back again, and now, with the brake pads off the wheels, I'm ready to start. A glance down the line shows that all the other propellers are also in motion, and the lead plane in *Staffel* 9 is already roaring off. The non-commissioned officer who oversees the take-off comes up and reports: "All ready!" A pull of the lever and all the hustle and bustle of the take-off area is swallowed up by the roar of our engine. The strange noise of the engine on take-off whips up one's nerves and senses again and again, so that they are in a state of extreme tension, until you settle down at 100 meters altitude in a peaceful lane with a sigh of relief in your comfortable, upholstered chair. That same sound puts one to sleep during long flights at greater heights. It smothers the loudest screams from your lips, so that the person you try to reach only understands the whispers with difficulty. It transforms the roaring, raging battle below into a mute pantomime for the spectator in the air, and yet a faint creaking and squeaking, a slight slackening of the engine will instantly rouse any pilot from the deepest "snooze."

For the first few hundred meters you keep going straight against the wind. First we crossed over the road, now we went over the railway to Brest in Lithuania, then we crossed over endless wasteland. Six minutes after take-off we got to almost 1,000 meters altitude. A look backwards shows the five other machines neatly staggered behind us. I take a slow turn, while Grote fires flare signal cartridges, and when we pass Kovel for the second time, this time at 1,600 meters, I can't help but admire the reliably good flying skill of the squadron, which was so flawless again, as they complete a few harmless leapfrog jumps over each other, which causes me to promptly knock my head against the observers' seat. Keeping in such a squadron formation is not as easy as it might look. It sounds so simple: "The leader goes in front and the others behind." But keep in mind that every crate works differently – one is fast, the other slow. One climbs well, the other not so much.

We're on time over Dubowa, and we set out on time. When the weather is clear, orientation is very easy, as you can always follow the train tracks. After half an hour, Maniewicze comes into view. Long before we get there, the anti-aircraft bursts greet us, but they are far away and much too low, as most of us have already gone higher than 3,000 meters. Only Noll and Kramer, who are crawling along at 2,400 meters, suffer not insignificant hits. We turn into the wind and I'm heading straight for the station. One or two trains rush East, a locomotive heads West, otherwise there's no other movement, and now the bombs rain down, 1, 2, 3, 4

– pull up; 1, 2, 3, 4 – pull up—turn--and away. Now the next gentleman, please! All together, the bombs deliver 600kg of the strongest explosives over buildings and facilities. When the weather is good, none of them miss. A shed is on fire, ammo trucks are blown up, tracks are often hit, and *Staffel* 9 has just came in to attack from the other side. As I fly away, I squint longingly at the Russian planes circling around the Maniewicze airfield. When we visit, they take-off each time to avoid getting caught in the shower of bombs, but they don't dare go higher than 200-300 meters, so an air battle is out of the question. If you attack them from above, they just land next to a flak battery, and then you're screwed.[28]

Since beginning the descent I throttled the engine down a bit and put the plane in a flat gliding position. Thus I slowly dropped lower at a relatively high speed. Above the front we are only 2000 meters up and we get a few greetings from the field guns, for which we promptly show our gratitude with machine gun fire. The Russian batteries have always stopped firing when we, recognizing their position by the muzzle flash, opened machine gun fire on them. At the front everything is calm. There are only isolated impacts – sometimes here, sometimes there – to show that there is any life at all in the labyrinth of the trenches. As we assemble over Mielnica, another field battery tries in vain to reach us. A kilometer or more away and shrapnel bursts explode ineffectively. Now it's time for yet another visit to an observation balloon, then a friendly greeting, a wave, and then off to home. Shortly before we get to Kovel, the big race starts and everyone wants to get there first. With a terrific ride – at 150, 160 km/hr (normal is 110 to 120), It's a few hundred meters above the city and the train station. A glide, flares, a few a few hops, clean landing.

The airfield is lively now. The mechanics take over the machines. Clean, clean, top up the fluids, look for bullet holes. When they find one, it is proudly displayed, as if they had been wounded in battle themselves. The hole is neatly glued, and a fine-painted inscription labels the place and day.

The observers gather up to the squadron leader and report in writing and verbally; the pilots still have work to do with the crates. They talk to the mechanics, the *Werkmeister* [master craftsman], and each other, and they examine and evaluate the barogram.[29] When everything is done, it's off to the cars and then home to the living compartments on the train, where we have the second and also very essential breakfast – now one certainly deserves it! – and we talk about everything again. It's terrific to talk about it all!

To the Western Front

End of January 1917
In this way, we kept very busy in Russia for a few months. Then we had some leisure time in Freiburg and Neubreisach until we were assigned to the borders of the empire.

Recently our *Geschwader* [combat wing] has a *Jagdstaffel* [fighter squadron]: von Haufen, Albert, Scheele, Hermann, Stern and I.

On January 22nd we finally had some flying weather. Scheele and I took off for a little tour of the front. At 4,000 meters we reached the front lines at St. Mihiel, flew along them, and crossed at Pont à Nommeny when two French Voisin aircraft[30] appeared halfway to the right, behind and below us. We made a wide turn into France, so that we pushed them over the front and, coming "out of the sun," dove down on them. Apparently they didn't notice us until we started shooting at close range. Unfortunately, we arrived with far too much speed; before we had fired a few shots we were already past them and had to turn around again. Flying ahead – at about 30 meters – I saw "my" French observer give me a salvo; at this great speed it naturally went far behind me.

The Frenchman twisted and turned in a steep glide and tried to escape over the front. We had fallen from 2,500 to 800 meters when my left gun jammed and he used that brief moment when I was trying to get it fixed to fly by me in the direction of France. I followed quickly and fired with determination using my right gun. At a height of 400 meters I saw the trenches darting past below me. To be on the safe side, I now tried my motor, which I had switched off while in a glide the whole time. Just my luck – it didn't start. Turn, curve around in a very flat glide homewards. At 300 meters I saw the French trenches below me; 200 meters later the foremost German line slowly, slowly emerged under me. Thank God! I'm not going to land on Joffre after all.[31] Moving up to the second line of trenches it was all open fields. If I tried to land there, I'd be immediately shot down. But just behind these lines there was a ravine. If I reached that, I'd be saved. At a height of no more than 5 meters I pulled over the first trees in the gorge – too bad, it was all forest with no possibility of landing. My poor 511 was a goner.[32] I let her hover, hover, and hover until she had just a little more speed, and then came a treetop that I couldn't avoid anymore. I dipped down in front of it and then pulled the machine up at an angle into the brush. After a wild splintering and cracking, there was a strong crash of my nose into the front of the body of the aircraft, and slowly my D 511 slid down from the tree onto the left wing, which crumpled like an accordion. I only had to unbuckle myself and extend my left arm, then I stood on my head and after some kicking around in my thick coverings I stood upright in a forest of bushes covered in deep snow. People from the improbably high house number 400 soon appeared out of the woods and took me to the command post of the battalion commander of the front line, *Rittmeister* [cavalry captain] *Freiherr* [baron] von der Goltz, and his adjutant, *Leutnant* Lustig, just 50 meters away. I was received extremely well by these two. I got field accommodations, but provisions were excellent and I got enthusiastic support with everything: dismantling my machine, transport, questioning of eyewitnesses, etc. With the testimony of five eyewitnesses in the front line who saw the Frenchman fall, the unit reported the kill. It would be quite nice if it were also recognized to be my kill – we'll have to wait and see. It's too bad that my brave D 511 has been completely used up in this movie.[33] Scheele got transferred to the "Boelcke" fighter squadron on January 27, 1917.[34] The lucky guy! *Vivant sequentes*[35]: for the time being I have the advantage of being able to take over his machine and not be put of action because of my bad luck. –

Mid-February 1917
Telegraphic inquiry to Richthofen, leader of *Staffel* 11.

Would you happen to need me?
  --Schäfer

Reply:
You have already been requested.
--Richthofen

Under *Freiherr* [Baron] von Richthofen

Early March 1917
With a real *Jagdstaffel* [fighter squadron] it's a whole different operation than with an imitation one.[36]

 On March 1st we took some new aircraft on a fighter patrol, which failed completely because we did not fly in a group and therefore each of us faced superior opponents. I struggled for a short while with four Sopwiths,[37] but couldn't really get on one for very long because the others were always breathing down my neck. In the afternoon we flew again, and this time the squadron flight worked, but there were no Englishmen. Things went better on the 4th [of March]. Our second flight had been out in the morning without doing anything. First flight[38] was getting ready. Richthofen wanted to go with us, but had to stay at home because of a small mechanical defect. So I took the lead. We were hardly at the front when an English squadron appeared and flew into the Loos area. We attacked them coming from Lens. I found out later that three Germans from *Staffel* Boelcke attacked at the same time. My first opponent escaped from me in a dive. Before I could follow him, I saw Almenröder[39] being harassed by two Englishmen and I took a deep breath for him. A Vickers single-seater[40] hit me from behind. I did a half loop and then let myself spin; two comrades who saw it thought I had been shot, as did the Vickers, who let me go unscathed. I wriggled myself out of the mess a bit so I could get a calm overview and then calmly took on a Sopwith two-seater, which started to burn after I put about 100 rounds into it. It dropped, rolled over and fluttered earthward in separate burning shreds, and I couldn't help but let out a loud cheer. Since I had a shot bracing wire, I took the shortest route home. In the afternoon we took off again, this time with Richthofen. Again we engaged with another squadron, and Richthofen attacked a Sopwith next to me and when he came out of the turn I immediately fired. Before I could get hold of him again, he passed through Richthofen's stream of fire and then tipped over unsteadily. After the battle we assembled north of Lens. Richthofen flew south with two others. I joined Wolff and Auslinger[41] towards a westward flying squadron and we attacked a Vickers single-seater, which began to smoke after a few shots, and then started to burn and crashed without stopping. As I watched him happily, a second Vickers shot at me from overhead. I immediately started to turn and after a turning battle [*Kurvenkampf*] that lasted for seven minutes, I finally got above him and shot him with a full burst. He smoked, stopped firing, and tried to get away by diving. Every time he picked himself up, I shot him from behind again.[42] So I pushed him down towards the ground. He wanted to land but couldn't find the space and he threw his crate into some trees on a country road. But I staggered home at an altitude of 200-300 meters. I was so far inside France, Béthune area, that it took me six minutes to get back to the front. The nice Englishmen expended a fortune in artillery ammunition on me without even scratching my crate. In this way I made a decent debut at my new club after all. The commander of aviation [*Kommandeur der Flieger*] even wants to put me in the army reports! My first kill is confirmed. He lies in our trenches. For the second I have two aerial witnesses and I'm hoping for confirmation; the third is just "quiet heroism," without witnesses, and I can't expect confirmation of him.[43] But in my private list of victories he certainly counts, you can count on that. He's right there, as he has a right to be. He's entitled to be on it.

 It snowed on March 5th, but today, the 6th, it's nice again. A fighter patrol in the morning yielded nothing. By noon, on the other hand, all hell had broken loose.[44] A large

Above: (image from original text) *"Leutnant Schaefer in seinem 'Albatros' "* -- *"*Leutnant* Schaefer in his 'Albatros'"* (colorized)

reconnaissance squadron wanted to cross the front with about 15 aircraft. Eight of us got to him. I came across a couple of Sopwiths and shot one in flames. A second flew up alongside me and kept shooting past me from behind. Now I locked on to him and shot off his rudder. I couldn't do more because others were coming at me from on my neck from behind and I had to dive away. I don't know what happened to him. On my way back I had a brief encounter with five Englishmen, whom I avoided after a short flight. They shot up Richthofen's engine, Lüttert got scratched.

Introduction to a Flying Manual

In my opinion, if you want to publish a pilot's manual, you shouldn't forget to write a dictionary for it. The language of pilots has too many expressions of its own that the layman can't understand. If he hears one pilot telling the other something, or if he only understands half of it, and if he reports in a way that circumvents and paraphrases the technical terms, then it's no longer a real pilot's story.

The population of pilots fall into two large groups: the pilots, nicknamed Emils, and the observers, known as Franzes – of course 'Franz' in the singular. An ominous angel – this is such a generalization that you can forgive me for the indictment – once defined the fundamental difference between the two with the question: "Tell me, dear, are you actually a pilot or do you just travel along with us?" Emil controls the plane, Franz sits at his mercy inside, delivered from life or death. The observer shows the way– in short, he "franzes."[45] Sometimes he doesn't pay attention either and he gets *verfranzt*.[46] Excited Franzes tend to want to be involved in piloting the aircraft in an advisory or even helping manner. If Emil's passive resistance is of no use, he punishes him with personal jabs [*Personalböen*], which are inflicted in the form of unexpected jerks in the steering. Another much loved solution includes Emil rocking Franz thoroughly back and forth. But I admit that the surest way to calm him down is by pumping the pressure line. We have a pump for the motor that creates air pressure in the fuel tank and thus provides fuel for the engine. Sometimes she doesn't do her duty, so the crew has to rotate the hand pumps. If you want to keep your Franz permanently and pleasantly occupied, you quietly and secretly open the valve on the pressure line until the pressure gauge shows the pressure at zero atmospheres, and the poor fellow, who is just along for the ride, has to pump, pump, pump, until Emil, overcome with pity, turns off the valve again. However, these cases are the exception. Franz and Emil are consistently good friends, as they depend on each other for better or worse. Their comradeship is strong through storms and fog, and through the dance of explosive bursts of enemy defense guns and in nerve-wracking, wild and wonderful air combat.[47]

The plane is called the barge or the crate.[48] The machine guns are shortened to M.G., machine gun fire is represented by a series of blasts. Those who know the language would like to call the "take-off" a "flight departure,"[49] but I hardly believe that the old, gray pilots can wean the small, big and really big aces from the word "take-off." What else should we call the take-off officer, take-off crews, take-off flare and take-off flag, the take-off log keeper and our cozy old take-off hut? These are at the very least verbal monsters. Before the take-off, the engine has to warm up for a few minutes and then the brakes are applied before it's tested briefly at full throttle. It's not allowed to bang, snort or puke.[50] When escaping, you first have to push the crate nose down, tail up, and get some momentum. If the motor revs up, that is, it slowly reaches its prescribed number of revolutions, you carefully and gradually lift it off the ground. That's the way old guys start. Young pilots, careless chickens, and stunt pilots rip the crate off the ground and climb almost vertically. This is known as the *Döberitzer* or cavalier take-off.[51] If you're not careful with this one, if you lose too much speed or get a gust of wind, you'll slip, fall on your face and produce a more or less complete crash. In the mess in the evenings you then have to pay the crash bottle and get to wear the Knight's Cross of the Order of Aircraft Destroyers.[52] In flight, you don't use an altitude control. Instead, you push down to go lower. Of course, when you push down, the machine goes faster with the help of gravity. So, if I'm in a hurry, for example if I want to get quickly on an opponent, I go full throttle and go at such a high speed monkey ride[53] that the bracing wires rattle. If I want to break away, which can also happen, I pull back on the throttle, turn the crate upside down and encourage the plane to go into a dive, as vertically as possible.[54] To climb you have to pull the barge up, especially if it is not built in such a way that it will leap out from under you.[55] You have to be careful not to over-pull, or you'll wobble and slip. This, when done involuntarily, is not one of the more pleasant things about life in a heavy aircraft, but it is a popular ruse in a lighter single-seat fighter if you want to get away from a superior opponent.

After the flight, many people say that the landing is "unfortunately" next. If you fail, you have to fly a lap of honor if you're not already lying in the mud. There are many types of landings: bumpy landings, forest and tree landings (rather unpopular), emergency landings that are often crash landings, wheel landings, tail landings, and egg landings.[56] You do the latter when you arrive home from a cargo or butter landing.[57] Bumpy landings usually result in a bent landing gear axle. Emergency landings often involve a headstand or a clean rollover.

Finally, I have a request for the reader. Please explain to all your acquaintances that German planes fly. We actually do fly, but everyone who doesn't fly themselves talks about how we drive.[58] What's special about driving or riding? Nothing. Every ship rides, every train rides, and our inflated competition, the zeppelins, float in the air and ride to London, every person can go ride on a street car for 10 Pfennig, even crap gets driven.[59]

But *we*, we fly like a bird in the air and fulfilled of mankind's old dream.

We fly!!

With Richthofen against the English

To start with, we were without Richthofen.

It was a dreary, unpromising looking morning. The first wing [*Geschwader* I] was ready to take off. We sit around in the take-off hut rather reluctantly. No messages came from the front. There were no flak clouds in the gray sky, and even the artillery fire that used to be constantly rumbling on our front had almost completely died down. In the end, it got so uncomfortable and boring that I suggested: gentlemen, we want to be open for business, we want to sneak up to the front, let our bones get shot up,[60] and if nothing's going on, we'll just fly home again. Then at least we'll have earned our second breakfast. Richthofen wanted to go, but his engine wasn't running properly, so he stayed at home and I was in the lead. There were four of us today and at first we flew straight to the front, which we reached near Arras. Far and wide, above and below, in the air and on the earth, one found the deepest peace. In leisurely curves and circles we approached Lens, flew over Lens and Loosbogen, and just as we aimed to set out for home, a shadow appeared near us in the haze. Two, three behind him, six, seven – a whole squadron of Englishmen. They approach from in front of us and want to intersect. We tighten our formation a bit and strike out against them as one unit. I had one well in my sights, but just as I was beginning to close in it veered away and I sped past, because I was going too fast. When I wanted to follow him, I saw that my friend, little Karl,[61] was not doing too well. He hung behind an enemy who fought back aggressively. Meanwhile, a second had settled under him and, like a rotten guy, he was shooting from below, and a third came straight at him from above. First I fired a burst into his face, so he jumped away terrified, then he got in the way of his comrade, so little Karl had some breathing space again for a while. But now those two joined up with my first opponent and took me on together. I now knew where to shoot first, and finally there was such a threatening rattle all

Above: (image from original text) *"Von Leutnant Schaefer brennend abgeschossen"* – "One shot down in flames by *Leutnant* von Schaefer."

around me that I preferred to disappear for a short time. So I pretended that I was shot down, pulled the machine up and let it spin down a few hundred meters. My opponents and even some of my comrades who saw it thought I'd been shot down. When I took control of the machine below again, I was alone. The combatants circled high above me. I flew around the fracas in a long turn, and just as I reached the right altitude again, reinforcements came: three planes from the Boelcke squadron. With them I attacked our opponent again and this time I got luckier. I caught a Sopwith who was relying on his observers' fire and kept slogging straight ahead.[62] That was easy aiming. After almost a hundred rounds, a large red flame erupted from the pilot's seat and the Englishman fell helplessly. His fall was the signal for a general retreat, and the squadron scattered in all directions, while we dominated the field of battle. Breakfast was honestly well-deserved.

A Double

In the morning it was quite hazy. Around 8am the second wing [*Geschwader* II] had made a long flight without having seen a single Englishman. At 10am the first wing [*Geschwader* I] took off. The front was empty, only a few individual captive balloons stood peacefully at a low altitude. But far, far over there, above the English airfields, life was lively. The Lords[63] could be seen circling overhead, individually and in small groups, but no one came towards us. We had all expected that Richthofen would now wait adamantly until they came, possibly for hours until the last drop of petrol, because the weather had become wonderful and the operation had to finally get started. To our great astonishment, our lead aircraft suddenly took a sharp turn and headed home. Maybe he had a faulty engine or didn't see the English over there. So I stayed on my course and two others remained with me. But Richthofen comes back, circles around us once and waves unequivocally: home. At the airfield he calls us together: "Gentlemen, we have ten minutes to fill up on fuel and ammo. In less than half an hour we'll be in a murderous operation[64] at the front, and I want to deploy more planes for this. Please brace yourself and get everything ready for when it starts."

It would soon become clear how right he was. Telephone reports were already coming in from artillery planes taking

Above: Famous Sanke card of pilots in von Richthofen's Jasta 11, Spring 1917. As marked on the card, Schaefer stands second from left.

off individually at Arras and Vimy, at Loos and Souchez. And now the expected news also arrived: a large English squadron crossed the front in the Loos salient. Half a minute later we were all in our planes and roared off. The explosive bursts of the anti-aircraft guns showed us the way. About six to eight kilometers on the German side, the squadrons met. Fifteen Englishmen were at an altitude of 2,000 meters, and we just reached 1,500 meters. Quietly, seemingly without being noticed, we passed under them to the West. Arriving at the front, now roughly level with the enemy, we turned around. If he wanted to go home now, he'd have to force his way through, and we'd cut him off. We were now able to take a calm look at the gentlemen. They were six fat Vickers, large lattice-tails,[65] photo-reconnaissance planes, and above them were nine Sopwith two-seaters,[66] light combat aircraft – the first attack had to be directed at them. We had slowly climbed above them and now we were each looking for an opponent. They didn't make that easy for us, as the English didn't fly straight for a second. They were constantly turning and circling and twirling around together; it was impossible to calmly take on a single one. One of them got separated about 500 meters away from the squadron, and Richthofen immediately took him with a rolling maneuver. I grabbed on to the last one of a group of three, sat behind him and fired a full burst. He squirmed and twisted, but I hung on close behind his tail and didn't let myself get shaken off. A second flew alongside me at close range and I saw the observer standing up and firing at me; but his tracer rounds left trails of sulphur all past me. When shooting from the side one has to give him a considerable lead at these enormous speeds.

I was now sitting so close behind one Englishman that the other at my side would have had to aim at his comrade to hit me. Because he didn't dare do this, he shot way behind me. So I let him shoot and calmly continued to deal with my original opponent, who finally burned and crashed. I had been waiting for that, so I'd have the chance to spoil the second man's shooting as quickly as possible. I rushed at him in a right turn, pulling my machine gun stream across his machine, and then with a quick left turn I was behind him. With my first burst, his rudder flew off and I saw him slide away. At the same moment, there's a rattling noise behind me. I turn around and see a large rotary motor within my reach, and over it a red muzzle flash spits long, thin threads of smoke at me. The first hits are already crashing into my plane; only tearing away helps. With a jerk I turn my crate upside down and it goes into a deep vertical nosedive. When I was below in peace and security and looked around again, I just saw one of the fat Vickers fire off a white flare, which means in German: the operation is abandoned. Thus I could fly home peacefully. In my report I put: "One Sopwith on fire, I shot the rudder off a second one, I don't know what happened," derisive, hellish laughter erupted: "The observer seems to have thrown his map at you,"[67] "The observer must have cried about your miserable shooting and his handkerchief flew off him." Those are the most harmless annoyances that I had to put up with. When there was no confirmation of the shooting until the afternoon, I was about to give up hope myself, but towards evening the Englishmen finally got himself found. Infantry Regiment X[68] sent a machine gun, the remains of a photographic apparatus, a piece of the wing and other rarities with the report that his plane had tumbled down out of control from the morning air battle and has broken up into its different parts over the trenches. The fuselage and one wing lay in the English lines, and the rest rained down on this side. So it was a double after all.[69]

The Massacre of Children[70] in Loosbogen

People must be lucky, and every good deed carries its own reward. These two beautiful sayings once came true for me. We had brought down all sorts of Englishmen lately, many shot dead, some captured alive. We made a neat list of the former, had a few words written to the parents of the latter and packed the whole thing in a field post [*Feldpost*] package. It was weighed down with a few pieces of coal to show those over there that there was still no shortage of coal, packed together and tied with a ribbon from the iron cross. With this improvised drop bag, I flew deep into the Loos salient from Lens one fine morning and threw it far behind the lines. I don't know if it was found, but we didn't get any confirmation of receipt.

A light morning mist hung in the air and was probably the reason why the flak shot at us so godawful poorly. It was all right with me, and happily, with the uplifting consciousness of a good deed done in my heart, I went home peacefully. I flew just over the bolt position,[71] which the English have created somewhat like the string of a bow. Up ahead the large canal to Douai came into view and flashed brightly in the morning sun. I saw a lonely plane circling overhead, a little lower than myself, not being shot at. I look and can't believe my eyes because I think I see a BE (British Experimental), an ancient, slow crate.[72] I press toward him and he doesn't take any notice of me, probably because I come from a place so far away from England. I am now within a few hundred meters and can clearly see the aircraft type and the national insignia. It is actually a BE. Franz [the observer] hangs halfway out of his compartment and keeps his eyes on the ground. Emil [the pilot] is probably dozing. Suddenly they see me. Startled, the observer jumps up and almost goes for his machine gun. The good old BE makes an improbably awkward turn and comes back at me! What audacity! I let him slip away from under me, turn around briefly on my hindquarters[73] and line him up cleanly right in front of me. After almost a hundred shots, his petrol tank explodes and he falls in a powerful plume of flames and smoke. The whole thing had delayed my way home by barely a minute, but now I'm hurrying up double-speed because I had good news to deliver. In addition, however, the anti-aircraft guns had gradually shot their way up towards me and they hurled their anger and disappointment in countless shrapnel and bombs. When I landed, I was greeted by my comrades congratulating me. All the haste had been in vain, as the anti-aircraft crews had already reported that a yellow Albatros single-seater with a black tail had shot down and Englishman in flames over the front. Unfortunately, we are not yet as fast as the electric waves in the telephone line.

How Richthofen Was Shot Down and Wounded

After a big squadron fight, during which each of us engaged at least two Englishmen, two of our planes were missing, including Richthofen. There was great dismay, frantic telephoning in all directions. Finally a message arrives: "*Leutnant*[74] von Richthofen, wounded in a dogfight, has made an emergency landing at Hénin-Liétard." As soon as possible I take Richthofen's chief aircraft crewman [*Flugzeugwarte*] in a car and drive off to look for him. In

Above: Sanke card of Leutnant Schaefer.

Hénin-Liétard, the usual big questions begin with: "Do you know anything about a German air officer who was shot down?" The local commander's clerk had heard that a German airman had landed and was seriously shot in the chest and arm. He didn't know where he was. Maybe in the field hospital? A military policeman [*Feldgendarm*] had seen the flying lieutenant [*Fliegerleutnant*] – a terrible word -- in the car of the infantry commander's office. There is no one who knows that the gentlemen are all at the mess for dinner. At the mess! After wandering about for a long time, I'm finally in a little side street. A friendly entrance, a nice anteroom, and next door there's the loud clamor of voices and the clinking of cutlery. How will I find Richthofen? I am announced, brought in and stand in front of a large group of people who have obviously dressed up for the official meal, and in the middle of it Richthofen is in his greasy old wool sweater[74] that he always wears when flying, still black in the face, but completely healthy and radiant. He has a plate full of oysters in front of him. He's too fond of them. For a dozen oysters I think he'll sell you one kill of an Englishmen. He wasn't shot either, "just" shot down. That is, his engine was shot up. The wounded man is the other missing gentleman of our squadron, who also crash-landed nearby and is now in the infirmary. Incidentally, his wounds were not as severe as initially reported, because after only one day he was flying again. The combination of these two emergency landings gave rise to the tale of Richthofen's injury.

As we drove to the landing site, Richthofen told me about his reception in Hénin Liétard, which he enjoyed very much.

"While I'm sitting in the plane, a car roars up and a completely distraught, agitated gentleman gets out. On request, he willingly offers to drive me to the next place where there's a telephone. I climb out of my crate and go with him to his car. 'So, you're completely unharmed, nothing happened to you?' he asks. As you can see, I'm healthy and whole.

'But' – and here the questioner stops, startled, turning to the plane – 'what's the matter with your vehicle's driver?'

Driver! The only thing missing is if he would have said chauffeur! I really don't know what he meant for a moment, until it dawned on me, and 'I drive alone' came out of my mouth as smoothly as I never thought it could. By the way, he was extremely kind and helped me in every way; but he could not at all grasp that this event, which interested and excited him so much, left the main participant completely cool. He asked me at least six times to calm down, and he wanted me to rest in his bed or on his sofa, and when I told him I'd already shot down 26 Englishman, he told me I shouldn't try to kid an old man.[75] But he gave me oysters to eat, and that made up for everything else."

With that, Richthofen climbed into the other plane that had made an emergency landing and was still flyable, and he shot down his 27th Englishman that same afternoon.

Bad Luck

The weather had been bad for a few days, so the Lords had been unable to undertake any long-range reconnaissance. Finally, the barometer started to rise, the weatherman[76] announced fine weather and a magnificent sunset gave hope for the best for the next day. We went to bed early, so that we had a good night's sleep, and as a good-night greeting we said: "Tomorrow it must rain English pilots' blood!" That sounds very bloodthirsty, but in reality it isn't meant so badly, because it's better if such a poor creature[77] lands with us safely than if he just arrives in tatters as a miserable piece of coal and ashes.

The next morning it was pretty nice – the first real day

of spring with lots of sun and white feathery clouds in the blue sky. No reconnaissance squadrons had been sighted or reported by 9:30am when Richthofen said they should have arrived. This is because he smells the English long before they arrive – one of the main reasons for his success. There's often nothing going on for hours. We sit around lazily in the take-off hut, the weather has gotten worse rather than better, and suddenly Richthofen stands up and says: "Now it's time." We take off and with absolute certainty we find the enemy in a short time. It was like that again today. Barely ten minutes after take-off, we had just reached the front when a squadron of six two-seater reconnaissance planes crossed over into our side. We were equally strong. The men couldn't have wished for a better and more sporting fighting engagement.[78] We calmly let them fly into our land as far as they wished; only when they turned West again did we attack. My opponent flew dead straight, and in my mind I saw him as number 17 on my kill list. I opened fire at 50 meters and then fired again and after ten shots: jammed. I quickly crawl under the enemy's tail and try to fix my gun; after a few shots it plays out the same way. Angry, I hang under the protective tail of the enemy and yank on the gun levers – no success, there's not going to be any shooting today. On the left an Englishman bursts into flames in a sudden explosion – that was Wolff's. To the right and above me, a second one slowly scorches and then falls, burning brightly – that was the little Richthofen, who, by the way, is a good margin taller than the "big" one.[79] The big brother also slides down with an already helpless opponent, and then there's a rattling and a crashing in my crate, which causes me to lose sight and hearing. In my eagerness to stay focused ahead I didn't pay attention to my protective roof above me. I didn't join in with an easy turn, and the friendly gentleman above me blew a handful of smoke trails into my face at a distance of 20 meters. With my stomach full of worries and an engine and machine full of hits, I returned to my airfield. All six Englishmen were shot down and I got none of them. Bad luck!

While the plane is being repaired, I pull another one out of the stable and try it out. Just when I'm done and fuel and ammo have to be refilled after the test flights and target shooting, the squadron takes off again. When it came back, there were two more Englishman lying on the ground, without me being there to help. Bad luck!

By midday more clouds had formed and the ground was covered in a thick haze. The three of us flew to the front – maybe we could "fish in the dark"[80] and take a careless artillery-spotting pilot by surprise. Two English fighter pilots, who were as gutsy as they were idiotic, probably had the same intention and suddenly attacked us. The one nearest to me suddenly noticed his mistake and immediately tried to reach a cloud while gliding steeply, but shortly before he could I caught him and had him perfectly in my machine gun stream for several seconds, after which he went down. Thinking it was a feint, I immediately went after him, but after an almost vertical dive of 1,500 meters I figured he was shot down, but in the prevailing haze I could neither see the impact nor find anyone who could have observed him from the ground. Bad luck!

Late in the evening we came up against another opponent. The three of us, Richthofen, Festner and I, attacked six fat Vickers over Lens.[81] One tumbled to the right, and the other to the left. I shot dead the observer of my aircraft and destroyed his engine. I had already forced him from 3,000 to 1,500 meters when I had a fatal gun jam again and had to watch passively as he glided flatly and escaped over the front. That was the worst luck I've ever had in a day. Of all days, this was the day that that the Richthofen squadron set their record of twelve kills.[82]

Jagdstaffel Richthofen, March 6, 1917
Dear parents!
Now you're going to finally get a detailed report about my recent health, because today is winter again. It's been snowing in thick, dense flakes for twelve hours. There's no thought of flying. We're gradually getting set up, so it's time to chat with you again. Well, first of all, I'm fine, perfect even. Ever since I've owned the iron cross first class, I've assumed a haughty, superior outward appearance and feel vastly superior to fellow human beings who are less decorated. I still have to think really hard about whether I should be in contact with you at all.[83]

In the first week that Richthofen was here, we had five kills, three by him. Yesterday, on the first flight day since then, there have been four kills – three by him, you think? Wrong! One by Richthofen and three by me! Yes, yes --Schäfer! Of course there was a lot of dumb luck as well as all sorts of bad luck. I will report more in the diary, but I have downed three, and one is in perfect condition, lying in the trenches. One crashed in flames, but it was so far deep into France that no eyewitnesses could be secured from the front, but two saw it from the air. I forced the third to the ground behind Béthune all by myself, with no witnesses from below or from the air. I hope that at least the first two will be confirmed. Werner Voss, who recently shot down his tenth Englishman, is also in the area; he understands the business but also goes into the attack like poison. Yesterday ten Englishman were shot down by the 10th Army alone and only one German made an emergency landing over there.

Staffel Richthofen, March 12, 1917.
Dear Mom!
Yesterday my sister Grete from Valenciennes called me just after having come home from the flight where I'd shot down my ninth. I flew there as quickly as possible and had a cozy coffee hour with Grete and our old acquaintances.
      Sincerely,
      Your boy

Staffel Richthofen, April 6, 1917.
Dear parents!
Happy Easter! Hopefully I'll find time for a longer report during the holidays.
    Numbers twelve and thirteen fell today, which are confirmed for me as eleven and twelve.[84] It's glorious spring weather here and it's been raining Englishman for the last eight days.
      Emil

Staffel Richthofen, April 9, 1917
Postcard
I'm doing very well
Wasn't the Easter telegram sent by *Extrablatt* [newspaper] a fine bit of news?
Today I've already got 14!
      Emil

Staffel Richthofen, April 10, 1917
Dear parents!
Unfortunately, we've been so busy, so writing was out of the question. Every free minute was spent sleeping. It was a little better yesterday and today. There is such a storm that flying is almost impossible. So we have some quiet time and yesterday we were able to drink a bottle to celebrate Richthofen's promotion to *Rittmeister* and my mention in the army report.
    Wasn't that a nice bit of news about my well-being? The day before yesterday, my number 13 fell, yesterday was number 14. The latter was unfortunately way beyond our lines, and given the mess that's overwhelming everything at the moment, we can't get confirmation, so I didn't register him until now. We're all very happy about Voss getting the *Pour le Mérite*, he really deserves it. He now has 24 properly credited kills.
    With my number eleven, Richthofen gave me a magnificent photo of himself with his own signature, which I am very proud of.
      Best regards,
      Emil

Staffel Richthofen, April 23, 1917.
Dear parents!
Since I wrote to you last eight days ago, the days have been rich in adventures. I've shot down four since then, but one landed fairly easily in a hollow where the artillery couldn't reach it, and in another situation a gentleman got involved and took one away from me in the melee. So on my official list it came out to only 20 and 21, though in reality I've got 25. During the last victory yesterday evening after 8pm over the lines, my engine was shot up by machine gunners on the ground at a very low altitude over Monchy, which caused me to land between the lines about 50 to 60 meters from the English trenches. In spite of all the shell holes, I landed without a hitch and crouched into a shell hole for the next two hours, where I waited until twilight, with the help of the cigar case that I happened to have in my pocket. As soon as it was possible, I began to march onwards, at first jumping out and scooting from shell hole to shell hole, eventually at a somewhat slower speed. The night was so dark, and I was so worried about English patrols that I slipped through all the folds in the terrain like a hunter, so that I crossed the first and second German lines without being noticed, but also without seeing more than a few people wearing steel helmets, whose nationality could not be determined, and whom I avoided just to be on the safe side. After almost an hour of walking, I arrived in what looked like a kind of trench, squeezed myself into a nearby crater and waited for people to start talking to each other. At first I thought I heard English, but it was Bavarian.[85] After another hour's march through an area polluted with gas shells, I came to Vis en Artois. There I found a field kitchen unit that took me with them. On four legs [horseback] we galloped through the burning village along the shell-torn road, which was still under fire for a good stretch, and after 1 ¼ hours, at a somewhat calmer pace towards the end of it, we reached Sailly. From there, at around 5am, I was able to communicate with my squadron, which had already given up on me.
    Today I lazed around all day despite the glorious flying weather and now I feel very good again. The infantry line into which I was received last night, was attacked by the English this morning. By the way, last night our squadron achieved its hundredth kill. Today we've already got a few more. Wolff caught up with me, but that doesn't bother me much. I'm glad that yesterday I came out of the story in one piece, and in the future I will survive the low altitude kills deep into their lines. Only by being calm can I achieve that.

Staffel Richthofen, April 26, 1917
Dear parents!
Today is a very eventful day that I'm celebrating with

Above: Pilots from *Jasta* 11 standing in front of von Richthofen in his Albatros D.III. Schaefer is just below von Richthofen.

very mixed feelings. His majesty the emperor deemed it worthy to bestow the Knight's Cross of the Royal House of Hohenzollern on *Leutnant* Schäfer. That's the purely enjoyable part. In addition, an order from the Commanding General of the Air Service states: "*Leutnant* Schäfer is appointed commander of *Jagdstaffel* 28." I find that fun. And I'm proud of the fact that I was immediately appointed leader, not just "acting leader." But I'm sorely sorry that I now have to leave the squadron that I love, the famous group of comrades, and above all leave Richthofen. If I had the choice, I would stay here a thousand times over. One consolation is that *Jagdstaffel* 28 is still flying against the English. Well,

hearty greetings, and if I don't write for a few days, it's because there's a lot to do. I probably won't be flying for a few days.

Emil

Telegram from the field, April 30, 1917
To the family of Emil Schäfer, Krefeld

Just received the Pour le Mérite!
Leader of *Jagdstaffel* 28,
Leutnant Schäfer

Royal Württemberg *Jagdstaffel* 28, May 4, 1917
Dear parents!
Many heartfelt thanks for your kind letters, for all the congratulations and other good intentions and for the many good or well-intentioned admonitions. Be assured that I approach every new opponent not only with an actual superiority gained through all past aerial combats, but also with a sense of calmness generated by my consciousness of this superiority. Calmness, cautiousness, and cold blood is everything and these promise even more success than the wildest daredevilry.[86] I am now faced with a task that I have never faced before. I hope and believe that I am completely up to this task. *Staffel* 28 is staffed with good pilots. Two Englishman have been shot down in the months before my arrival. In the four days that I've been leader, we've sent down four – the prospects are good.

Now to answer a few of your questions: all air service units are set up so that parents are notified in the event of an accident.

The phrase "no news is good news" is always correct.

This is the first quiet hour that I've had for myself in a week. All mail is still stored at *Staffel* Richthofen on my instructions and will not be sent here for the time being, because I just don't have the time to read it or even to answer it. In the month of April I received 101 items of mail and replied to 70. This is to increasing every day – burdens of fame!

Your boy

Jagdstaffel 28, May 10, 1917
Dear parents!
I really enjoyed the reports about Krefeld's interest in my successes. Please send me the extra news pages that appear, and they'll be posted on the wall like the enemy aircraft serial numbers.[87] This makes for a magnificent hall of heroes décor.

Of course, leave is out of the question now. First, the *Staffel* has to get into the swing of things here. You can't start your new job right away by asking your boss for leave. In addition, this place is a vacation compared to operations at *Staffel* Richthofen. Yesterday, I shot down another one, number 26. In the past, three kills in fourteen days would have made me feel like a very lazy, listless dog, but around here, one is happy just to see one Englishman in three flights.

Nowadays my squadron is shooting down Englishmen all over the place. Whoever would have told me six weeks ago that I could have calmly watched this happen, I would have laughed at them.

I expect a visit from our Grete any day.

Emil

Jagdstaffel 28, May 29, 1917
Dear parents!
It was a pity that our farewell in Bonn was so spoiled by the stupid cab driver. But now you've seen me fly once. I saw you clearly standing in the market square and waving. I got from Cologne to Lille perfectly in two hours without a stopover. The weather here is unfavorable for flying. We stand on the field all day and wait for a blue hole in the sky, mostly in vain. My invitation to the 19th Corps went very well. Today my cousin Ernst Hehner, who is in the area and in a quiet part of the trenches right now, visited me. Borchers, Höhne and Schramme, the old friends of *Jäger* 10, who are now at Reserve-*Jäger* 2, were also here recently. It was great to see old comrades again. Werner Voss arrived yesterday by air. He looked very good and wants to fly to Krefeld in the next few days.

With best regards,
Your Emil

Jagdstaffel 28, June 4, 1917
Dear parents!
I'm doing well. The enemy's operations are very strong, but we have so many fighter squadrons here and they are deployed so systematically that we only need to fly two or three times at most. There is no comparison with the stress of the Arras offensive. Cousin Otto Remkes is here right now and will be lounging in our quarters for a few days where he will be well-fed. He looks better than he did in Brussels a while ago.

I shot down number 30 today.
Sincere greetings,
Emil

Service Telegram
June 5, 1917
To the family of Emil Schäfer, Krefeld

While at the head of his *Jagdstaffel* our leader *Leutnant* Schäfer fell this afternoon in an aerial battle against an English squadron.[88]
Jagdstaffel 28

Endnotes

1 Schaefer use the term "*Jäger*," the German term for light infantry, which literally means "hunter." This term of course appears in the title of Schaefer's memoir, *Vom Jäger zum Flieger* – there and in other parts of his memoir he makes a play on the dual meaning of "Jäger." I've translated this word as "infantryman" or "infantry," and in

many cases also included the word "*Jäger,*" but the layers of meaning in the original German should be kept in mind.
2. "*Vive la guerre*" and "*A bas la guerre*" – "Long live war" and "down with war". There's a popular image of near universal enthusiasm for the outbreak of the war in 1914, as many groups welcomed what they saw as an inevitable, even desirable opportunity to show patriotic fervor and distract from political conflicts that had divided European societies for decades. However, evidence shows that it was a much more complex picture, as anti-war demonstrations were largely suppressed by conservatives (supported by the middle and upper classes) in power. In contrast, many in the socialist parties largely protested the war. See, for example Jeffrey Verhey's *The Spirit of 1914: Militarism, Myth and Mobilization in Germany* (Cambridge: Cambridge University Press, 2006).
3. "The Matin" is a famous café in Paris.
4. "*sale Allemand*" and "*ehlen d'Allemand*" – "dirty German" and "German dog."
5. The *Paris-Midi* was a popular newspaper.
6. Interestingly, he uses the word "*Dächse*" – literally "badgers" – in context, this seems to be slang for rifles.
7. The "*Damenweg*" was known as the "*Chemin des Damens*" – both mean "Ladies' Path." It was a major road near Soissons, and fierce fighting took place there during the First Battle of the Aisne in September 1914.
8. "*Tschako*" – a visored helmet worn by the Prussian *Jäger* battalions.
9. "Tommy Atkins" was a nickname used by German soldiers for the British.
10. "*Hatte ich doch mein Fett weg....*" – slang: literally, "Actually I had my fat removed," which I've translated to the phrase familiar to Anglo-American readers, "Actually, I had a close shave."
11. This is the the famous soldiers' song "*Ich hatt' einen Kameraden.*"
12. He's likely referring to the devastating First Battle of the Marne and the Battle of the Aisne, in September 1914.
13. Joseph Glidden was the American inventor of barbed wire, which was used for cattle farms, in the late 19th century.
14. He used the Swabian slang term for "head": "*Däz*" – thus the translation to "noggin."
15. Here he's referring to the famous *Pickelhaube*, the infantryman's leather helmet with a spike used during this early phase of the war.
16. "*Schinkenklopfen*" – "beat the ham": this was a popular game, involving spanking, that goes back to the Middle Ages.

Above: (image from original text) *"Leutnant Schaefer mit seinem roten Flugzeug in der Heimat"* – *"Leutnant* Schaefer with his red airplane in the homeland."

17. Here he's joking about what the men look like when dressed in make-shift towels and covers.
18. Schaefer signs off as "Your *Filius*" – "*Filius*" is Latin for "son".
19. "This time as Franz" means "this time as observer" (Franz was the German nickname for an observer, Emil was the nickname for the pilot). It looks like Schaefer's observer, Peter Martin Lampel, signed this postcard under Schaefer's joke to his parents. See Lance Bronnenkant, *Blue Max Airmen*, volume 7, p. 43 for clarification.
20. Kovel is presently in Ukraine.
21. LDD refers to "Leicht Doppeldecker" or "light biplane", a general term for single-engined biplanes. C674 is the aircraft's serial number.
22. In the original it says "B.U.K.", but this may be a misprint. It would most logically be the K.U.K, a reference to the Austrian military who shared their airfield.

23 Schaefer's description of Russian troops here reflects widespread cultural perceptions held by many Germans that the Russian people were culturally and racially inferior.
24 Turkestan was part of the Russian Empire.
25 It's not clear why Schaefer puts "town" in quotes when he describes Kovel. It had a relatively large population as it was a transportation hub in 1916. Perhaps after the Battle of Kovel in 1915, in which the town suffered considerable damage and the Russian army took heavy losses, it no longer seemed so substantial.
26 "*Kahn*" – barge or boat – a nickname that pilots often used for the larger and slower two-seater aircraft.
27 A Hungerian Jucker is a large, bulky type of horse that usually pulls wagons, in stark contrast to the relatively sleek Rumpler two-seaters. The "big AEG" is likely a reference to the AEG G-type twin-engine bomber aircraft.
28 "*…und man ist geliefert.*" – translating this to "and then you're screwed" is a bit colloquial, but in this context "*geliefert*" is slang for "sunk", "out of luck", or "screwed."
29 A barogram is the recording of atmosphere pressure, usually on a sheet in a linear graph.
30 Schaefer calls this French aircraft a 'Voisin D' – a 'D' would be the German designation for a single-seat biplane fighter, but 'D' wouldn't be used to designate a French aircraft. It's not clear what type of Voisin this was.
31 General Joffre was the commander-in-chief of the French army until the end of 1916.
32 Here he's referring to the serial number of his Albatros D.II 511/16
33 Interestingly, he refers to his experience as a 'Film' – a movie – like it were an out of body experience.
34 The "Boelcke" fighter squadron was *Jagdstaffel* 2, named after its commander, who was killed in action in October 1916.
35 This Latin phrase would have been familiar to an educated audience – it means "long live those who come after."
36 "*Nachahmung*" – I translated this to be "imitation", but more colloquially one might say a "knock-off."
37 Since it's March 1917, he's likely referring to Sopwith Pups.
38 He refers here to "*Schwarm I*" (literally "Swarm I"), and in the previous sentence to "*Schwarm II*". I've translated these to first flight and second flight.
39 Strangely, Schaefer misspells the name of Karl Allmenröder, a famous ace in *Jasta* 11 who was killed in action in June 1917.
40 He's likely referring here to the British DH.2 fighter – German pilots regularly called any pusher type British aircraft a 'Vickers'.
41 Here Schaefer refers to the famous ace Kurt Wolff. However, his reference to "Auslinger" is confusing. He likely means Franz Anslinger, who would transfer to *Jasta* 35 a few days after the events of this narrative.
42 "*In den Nacken*" – literally "in the neck."
43 Pilots needed confirmation by eyewitnesses for their victories to count as a confirmed kill.
44 "*Gegen Mittag war dagegen der Deibel los.*" – "*Deibel*" is slang for "devil", but I translated this into the more Anglo-American colloquialism, "all hell had broken loose."
45 "Here Schaefer has turned "*Franz*" into a verb – "*franzen.*"
46 "*Verfranzt*" -- Schaefer is jokingly equating 'getting lost' with being a 'Franz' – i.e. "he franzed", to mean "he got lost"…..clearly a disparaging comment from the biased view of an Emil like Schaefer – who, coincidently, went by his middle name, "Emil."
47 Intersting language here: "*…nervenpeitschenden, wilden, herrlichen Luftkampf.*"
48 "*Der Kahn oder die Kiste.*"
49 He distinguishes between the words "*Start*" and "*Abflug*" – both of which can mean "take-off. But there's a subtle difference, as "*Start*" is typically known in aviation circles as "take off", while "*Abflug*" is literally "fly away", or "departure."
50 "*Hierbei darf er weder knallen, noch runksen, noch kotzen.*"
51 "*Döberitzer*" – one from Döberitz. "*Kavalier*" – I've translated to "cavalier", which can mean a gentleman in arms, or someone with a flippant attitude.
52 "*Ritterkreuz des Ordens der Flugzeugzerstörer*" – an in-joke about the knight's cross medal.
53 He uses the term "*Affenfahrt,*" which literally means "monkey ride," a term found in other memoirs. I added "high speed" before this interesting colloquialism to help readers unfamiliar with this term.
54 "*…empfehle mich im Sturzflug*" – literally "recommend myself into a dive." I translated it to "encourage" to make it a bit more idiomatic.
55 "*…dass er einem unterm Arm wegsteigt*" – I translated to the more idiomatic "leap out from under you."
56 "*Eierlandungen*" -- literally "egg landings"
57 "*Güter- oder Butterlandung*" – the words are easily translated, but the meaning is a bit obscure.
58 He distinguishes between "*fliegen*" (fly) and "*fahren*" (drive, or ride).
59 "*…sogar Mist wird gefahren*" he's referring to carts that haul manure.
60 Some interesting slang here: "*…uns einmal die Knochen ordentlich durchpusten lassen.*"

61 "*Karlchen*" – likely Karl Allmenröder, a comrade of Schaefer's in *Jasta* 11. Allmenröder was killed on June 27, 1917, just a few weeks after Schaefer.
62 A Sopwith with an observer – this was very likely the Sopwith 1½ Strutter that was his second victory on March 4, 1917.
63 Here he uses the English word "Lords", making fun of the term for upper-class Englishmen.
64 Interesting term here: "*Mordoperation*."
65 "Vickers" was a generic term used by German pilots for pusher aircraft, also known by the Germans as "lattice-tails" (*Gitterschwänze*). Since these were observation aircraft, he was probably encountering the F.E.2b.
66 Sopwith 1½ Strutters.
67 "*Der Beobachter hat wohl mit der Karte nach dir geworfen.*" "*Karte*" could be "card", but it is likely "map" in this context, and it seems to be a joke about the enemy plane losing its rudder.
68 Schaefer doesn't give the regiment number – likely a wartime security consideration.
69 Schäfter shot down these two Sopwith 1 ½ Strutters on March 6, 1917.
70 "*Kindermord*" – "the murder" or "massacre of chidren"
71 "*Riegelstellung*" – a "bolt position" is a defensive position often reinforced with bunkers or other emplacements.
72 He's referring to a B.E.2 British two-seater observation plane, which was obsolete and easy prey.
73 "*Hinterhand*" – "hindquarters" – interesting that he uses this term for flying.
74 Manfred von Richthofen was not promoted to *Rittmeister* (cavalry captain) until April 1917, a few weeks after this.
75 This would be March 17, 1917, judging from his statement in the next paragraph that it's the day of von Richthofen's 27th kill. By this date, Schaefer had eight confirmed kills.
76 "*Wetterfroch*" -- literally means "weather frog" in German, a colloquialism for the weather man.
77 "*Luder*" – I translated this as "creature" but in different contexts it's slang for something more pejorative.
78 "…*sportmässigeres Kampfverhältnis*" – Schaefer, and many other pilots, often depicted combat as a kind of sport or a hunt.
79 Here he's referring to Lothar von Richthofen as the "little" brother to the "big" ace, his brother Manfred. Lothar was substantially taller than his brother.
80 "…*im Trüben fischen*" – more literally, "fish in troubled waters."
81 Most likely these were F.E.2b two-seater pusher aircraft, often generically called "Vickers" by German pilots, as mentioned earlier.

82 This was likely on April 13, 1917, when Festner and von Richthofen both shot down F.E.2bs. During that month, widely known by the British as 'Bloody April,' von Richthofen's *Jasta* 11 shot down 89 aircraft.
83 Schaefer is of course joking to his parents here, with sarcasm dripping from his pen.
84 Though German pilots could technically only count kills confirmed by eyewitnesses, Schaefer kept kept an internal count of the total number of planes he'd shot down, including those that were unconfirmed.
85 This is an interesting little joke: Schaefer is originally from Krefeld, which is in Rhine-Westphalia. A Bavarian dialect would have sounded very different compared to his dialect, and indeed many Germans joke that Bavarian is so distinct that it almost sounds like another language.
86 "*Draufgängertum*" – recklessness, bravado.
87 Krefeld was his home town, and the town newspaper published stories about Schaefer in their *Extrablatt*.
88 Schaefer was killed in a dogfight with British pilots of RFC No. 20 Squadron on June 5, 1917.

Above: Emil Schaefer; the *Pour le Mérite* was added to the photo posthumously. (photo via Lance Bronnenkant).

Lt. Willy Rosenstein Fokker D.VII(Alb), Jasta 40

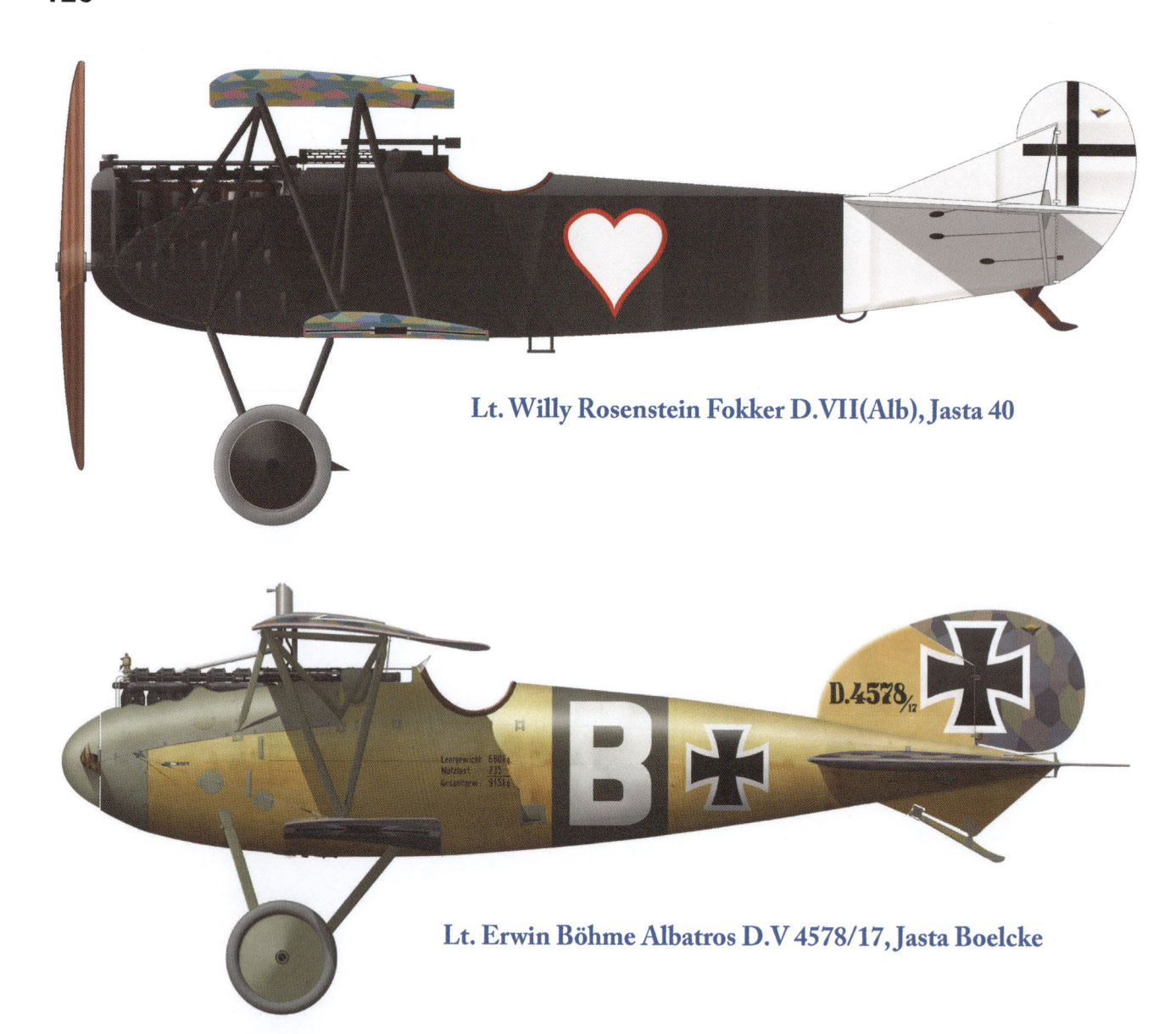

Lt. Erwin Böhme Albatros D.V 4578/17, Jasta Boelcke

Lt. Emil Schaefer Albatros D.III, Jasta 28

Printed in Great Britain
by Amazon